AF595053

Brain Asymmetry in Evolution

Brain Asymmetry in Evolution

Editor

Kazuhiko Sawada

MDPI • Basel • Beijing • Wuhan • Barcelona • Belgrade • Manchester • Tokyo • Cluj • Tianjin

Editor
Kazuhiko Sawada
Tsukuba International University
Japan

Editorial Office
MDPI
St. Alban-Anlage 66
4052 Basel, Switzerland

This is a reprint of articles from the Special Issue published online in the open access journal *Symmetry* (ISSN 2073-8994) (available at: https://www.mdpi.com/journal/symmetry/special_issues/Brain_Asymmetry_Evolution).

For citation purposes, cite each article independently as indicated on the article page online and as indicated below:

LastName, A.A.; LastName, B.B.; LastName, C.C. Article Title. *Journal Name* **Year**, *Volume Number*, Page Range.

ISBN 978-3-0365-0612-8 (Hbk)
ISBN 978-3-0365-0613-5 (PDF)

Contents

About the Editor

Kazuhiko Sawada

Education:

- Department of Biology, Faculty of Science, Yokohama City University (April 1985 to March 1989)
- Master Course in Regulation Biology, Graduated School of Science and Engineering, Saitama University (April 1989 to March 1991)
Received M.S. ("Rigaku Shushi", literal translation is Master of Science) in March 1991
- Doctorate Course in Biological and Environmental Science, Graduated from School of Science and Engineering, Saitama University (April 1991 to March 1994). Received Ph.D. ("Hakushi (Rigaku))", literal translation is Doctor of Science) in March 1994 from Saitama University

Occupation:

- Researcher, Department of Toxicology and Pathology, Nippon Roche Research Center (April 1994 to January 1996)
- Assistant Professor, Department of Anatomy, University of Tokushima School of Medicine (February 1996 to July 2001)
- Associate Professor, Department of Anatomy and Developmental Neurobiology, University of Tokushima Graduate School Institute of Health Biosciences (July 2001 to March 2007)
- Professor, Department of Physical Therapy, Faculty of Medical and Health Sciences, Tsukuba International University (April 2007 to March 2014)
- Professor, Department of Nutrition, Faculty of Medical and Health Sciences, Tsukuba International University (April 2014 to present)

Preface to "Brain Asymmetry in Evolution"

This Special Issue entitled "Brain Asymmetry in Evolution" comprises five papers published in *Symmetry*, and includes one review article and four original articles investigating unique aspects of the lateralized morphology and function of the brain in various vertebrate species, such as humans, non-human primates, carnivores, and zebrafish. A review by Sawada overviews the evolution of laterality of the cerebral cortical morphology in primate species with views of sulcal development. Original articles by Eckert et al. and Horiuchi-Hirose et al. investigate cerebral asymmetry in human associated higher cortical functions and arithmetic task performances and languages, respectively. Sawada et al. describe, in detail, asymmetric features of the cerebellar lobular development in ferrets. Petrazzini et al. report on the involvement of brain lateralization in the side biases of binary choice for two alternative stimuli in zebrafish. I believe that these findings are the key to elucidating brain evolution in vertebrates. I hope that this Special Issue will provide readers with new insights into understanding brain lateralization throughout its evolutionary trajectory.

Kazuhiko Sawada
Editor

Review

Cerebral Sulcal Asymmetry in Macaque Monkeys

Kazuhiko Sawada

Department of Nutrition, Faculty of Medical and Health Sciences, Tsukuba International University, Tsuchiura, Ibaraki 300-0051, Japan; k-sawada@tius.ac.jp; Tel.: +81-29-883-6032

Received: 24 August 2020; Accepted: 11 September 2020; Published: 14 September 2020

Abstract: The asymmetry of the cerebral sulcal morphology is particularly obvious in higher primates. The sulcal asymmetry in macaque monkeys, a genus of the Old World monkeys, in our previous studies and others is summarized, and its evolutionary significance is speculated. Cynomolgus macaques displayed fetal sulcation and gyration symmetrically, and the sulcal asymmetry appeared after adolescence. Population-level rightward asymmetry was revealed in the length of arcuate sulcus (ars) and the surface area of superior temporal sulcus (sts) in adult macaques. When compared to other nonhuman primates, the superior postcentral sulcus (spcs) was left-lateralized in chimpanzees, opposite of the direction of asymmetry in the ars, anatomically-identical to the spcs, in macaques. This may be associated with handedness: either right-handedness in chimpanzees or left-handedness/ambidexterity in macaques. The rightward asymmetry in the sts surface area was seen in macaques, and it was similar to humans. However, no left/right side differences were identified in the sts morphology among great apes, which suggests the evolutionary discontinuity of the sts asymmetry. The diversity of the cortical lateralization among primate species suggests that the sulcal asymmetry reflects the species-related specialization of the cortical morphology and function, which is facilitated by evolutionary expansion in higher primates.

Keywords: non-human primate; Old World monkey; evolution; evolutionary expansion; gyrification

1. Introduction

Brain lateralization has been reported in many species including vertebrates and invertebrates [1]. The gyrencephalic surface morphology is a distinctive feature of the cerebral cortex seen in some mammalian species including humans [2], nonhuman primates [3–5], carnivores [3,6–9], artiodactyla [3,10] and cetaceans [11,12]. The cerebral sulcal morphology in primates is largely influenced by genetic factors that manifest during the second-trimester equivalent [13,14], and is modified under the influences of developmental and environmental factors during the third-trimester equivalent to form adult features [15,16]. In particular, the cortical expansion, known as one of the developmental factors, is closely correlated with gyrification across primate species [17,18] and is involved in species-related patterns of sexual dimorphism and/or individual variability of the sulcal morphology [19,20].

Cerebral sulci are highly lateralized in humans. Leftward sulcal asymmetry reported in the central sulcus (cs) is associated with handedness [21,22]. Additionally, leftward sulcal asymmetry in the paracingulate sulcus has been linked with cognition [23,24], while rightward sulcal asymmetry in the superior temporal sulcus (sts) is relevant to fMRI-defined right-lateralized voice selective responses [25,26]. Cortical lateralization, along with sulcal asymmetry, has been classically considered to be a unique characteristic in humans [27]. However, sulcal asymmetry has been observed in nonhuman primates such as great apes [28,29] and macaque monkeys [29]. This report summarizes the results of our studies and other findings regarding the laterality of sulcal infolding in the cerebral cortices of macaque monkeys—a genus of the Old World monkeys. The pattern of sulcal asymmetry in macaques to other primates was compared to speculate on the evolutionary significance of the sulcal asymmetry in primates. The sulcal asymmetry of the macaque brain and its related characteristics are summarized in Figure 1 in comparison to other primate species.

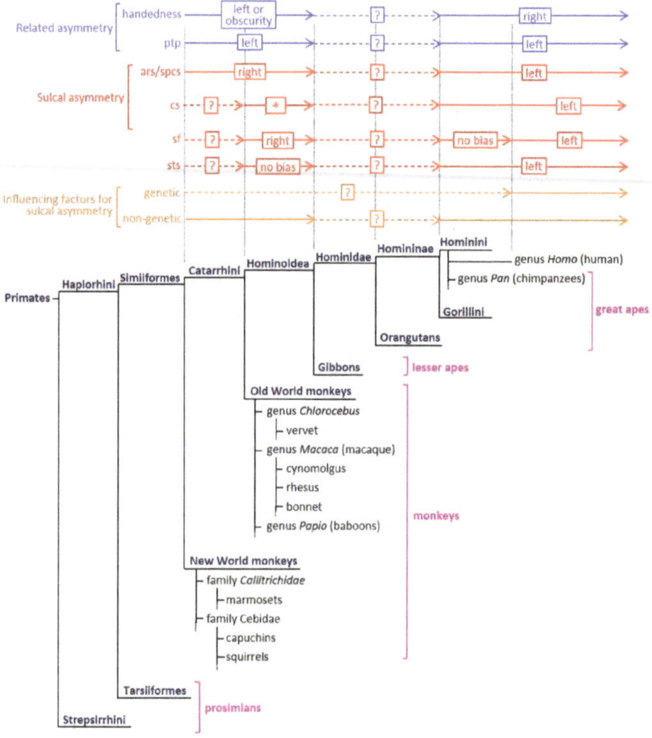

Figure 1. The evolution of sulcal asymmetry and its related asymmetrical characteristics among primate species. The sulcal asymmetry, its related asymmetrical characteristics, and influencing factors for sulcal asymmetry are summarized in reference to previous reports [4,5,13,20,28–47]. An asterisk (*) indicates that the cs symmetry is inconsistent in the Old World monkeys. The presence or absence and the direction of cs asymmetry vary depending on the species. ars, arcuate sulcus; cs, central sulcus; ptp, planum temporale; sf, Sylvian fissure; spcs, superior prefrontal sulcus; sts, superior temporal sulcus.

2. Asymmetrical Development of Cerebral Sulci

Sulcal morphology is influenced by genetic, developmental, and environmental factors [15,30]. In humans, the cs and sts emerge earlier on the right side compared to the left side during fetal development [2,31], which suggests a large influence of genetic factors on the asymmetric sulcal emergence. On the other hand, the sulcal length and depth were variable after birth in the baboon, a genus of the Old World monkeys [13]. Furthermore, sulcal infolding in multimodal association cortices was highly correlated with cortical expansion in cynomolgus macaques [16]. These findings suggest the possibility that the sulcal morphology in the Old World monkeys is modified after birth under the influences of developmental and environmental factors. Cynomolgus macaques experienced fetal sulcation and gyration symmetrically [4,32] and showed the sex-related appearance of sulcal asymmetry after adolescence [20,33]. Thus, sulcal asymmetry in the Old World monkeys may be largely attributed to environmental and developmental factors rather than genetic factors.

Although the New World monkeys are secondarily lissencephalic with reductions in body and brain sizes compared to their gyrencephalic ancestors [48], several primary sulci are sustained [5].

The cs, sts, and the parietooccipital sulcus emerged in adult common marmosets as small indentations, separately, on the left and right sides [5]. The incidence of each sulcus emerging was not biased between the left and light cerebral hemispheres [5], thus suggesting the individual variability of the emergence of asymmetric sulci. Therefore, the shift of nongenetic factor-related sulcal asymmetry from individual to population-level variabilities may appear before the split between the New World monkeys and Old World monkeys. Furthermore, the genetic factor-related sulcal asymmetry may occur after the split between the Old World monkeys and humans.

3. Sulcal Asymmetry in Macaques

The sulcal length asymmetry in macaques has been reported for more than three decades [49–53]. However, these studies failed to obtain consistent results. One possible reason for these discrepancies is that several of these studies used mixed-sex and/or -species samples for measurements of the sulcal lengths. After separating male and female cynomolgus macaques, we measured the sulcal lengths of left and right cerebral hemispheres and found the right-lateralization of the arcuate sulcus (ars) length in males, but not in females [20], following puberty [33]. The AQ values of eight representative primary sulci in our previous report [20] were analyzed using a one-sample t-test to determine significant-population level asymmetry of the sulcal lengths in cynomolgus macaques. A significant right-bias of AQ values (0.053 ± 0.069) was only identified after examining the ars length in adult males ($p < 0.05$). A sulcal map of the left and right cerebral hemispheres in cynomolgus macaques is shown in Figure 2. The ars in macaques is infolded in the lateral frontal cortex and identified as the early-emerging sulcal groups, which emerged in cynomolgus macaques between embryonic days 70 to 120 of the 160 day-long gestational period [4,32]. The cs, sts, and cingulate sulcus also appeared during this time [4,32]. However, the ars morphology in the Old World monkeys was greatly influenced by environmental factors rather than genetic factors, unlike the morphologies of other early-emerging sulci [13].

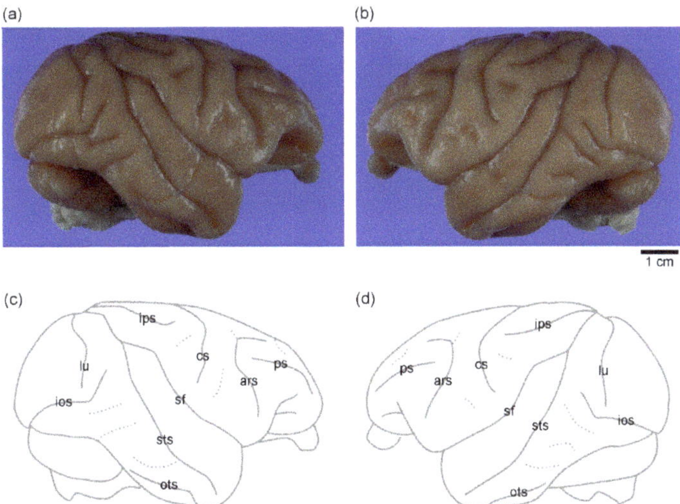

Figure 2. Eternal surface and sulcal map of the left and right cerebral cortex in cynomolgus macaques. (**a**) Photograph of the right hemisphere. (**b**) Photograph of the left hemisphere. (**c**) Sulcal map of the right hemisphere. (**d**) Sulcal map of the left hemisphere. ars, arcuate sulcus; cs, central sulcus; ios, inferior occipital sulcus; ips, intraparietal sulcus; lu, lunate sulcus; ots, occipitotemporal sulcus; ps, principal sulcus; sf, Sylvian fissure; sts, superior temporal sulcus.

One major limitation of our previous study (Imai et al. (2011) [20]) is the use of the "cotton thread" to measure the sulcal length. This method involves placing the cotton thread directly onto the cerebral surface to measure the lengths of each sulcus. However, this method cannot be used to obtain other information besides the sulcal length, e.g., the sulcal depths and surface areas. The sulcal depth, rather than the sulcal length, is known to be a susceptible mediator for evaluating the contributions of environmental factors [13]. Bogart et al. (2012) [29] conducted an MRI-based morphometrical analysis of primary sulci of macaques. This study used combined data from mixed-sex samples of two macaque species (rhesus and bonnet macaques), because no significant main effects or interactions were identified between either sex or species. Significant population-level asymmetries were obtained by the right-side bias in the sts surface area, but not in depths of any primary sulci. On the other hand, a significant individual asymmetry was revealed by the right-lateralization of the ars depth in macaques [29].

The ars in macaques is infolded in the lateral frontal cortex and putatively has cortical connections to premotor areas associated with arm and laryngeal representations, frontal eye fields, and somatomotor/orofacial areas [54]. Although the cerebral cortex continuously expanded after the completion of primary sulcogyrogenesis in cynomolgus macaques [18] and other primates [5,55], the progression of ars infolding correlated with the cortical expansion [16]. This phenomenon was similar to the late-emerging sulcal groups infolded in multimodal association cortices among cynomolgus macaques [16]. Therefore, the right-lateralization of the ars length and depth may be attributed to factors such as androgens, which are predominantly expressed in postpubertal males. This hypothesis is supported by a previous study showing the right-lateralization of androgen receptor levels in the frontal lobe in male rhesus monkeys, but not in female monkeys [56].

The sts is another sulcus that exhibits significant population-level asymmetries among macaques. A significant rightward asymmetry was revealed in the sts surface area [29]. Asymmetric pattern of the sts length may vary regionally among Old World monkeys, which consisted of three species including rhesus macaques: significantly right-lateralized at the lateral portion, but not at the medial and insular portions of the sts [57]. Intriguingly, the rightward sts asymmetry has not been observed in other families of nonhuman primates such as great apes (including chimpanzees) and New World monkeys (including capuchins and squirrels) [57]. In humans, the sts depth exhibited a significant rightward asymmetry that was relevant to fMRI-defined, right-lateralized voice-selective responses [25,26]. Asymmetric aspects of the sts morphology may be comparable between the Old World monkeys and humans. However, no left/right side differences were identified in the sts morphology among great apes, which suggests the evolutionary discontinuity of the sts asymmetry.

4. Comparison to Other Nonhuman Primates

Hopkins and colleagues reported sulcal asymmetry in great apes. They used MRI data from captive chimpanzees (*Pan troglodytes*) of both sexes to analyze the sulcal asymmetry due to the lack of sex differences in either the sulcal surface area or sulcal depth. The sulcal surface areas and/or sulcal depth were significantly left-lateralized in the fronto-orbital sulcus, superior precentral sulcus (spcs), sf, intraparietal sulcus, inferior postcentral sulcus, and the lunate sulcus [29]. The spcs in chimpanzees comprises part of the anterior border of the premotor cortex, and it is anatomically identical to the ars in macaques [34]. Interestingly, the surface areas and depth of the spcs were left-lateralized in chimpanzees—opposite to the direction of the symmetries in ars length and depth in macaques (right-lateralized) [29]. The precentral corticospinal system was associated with handedness in chimpanzees as well as humans [58]. Population-level right-handedness (preference for the right hand) was observed in both wild and captive chimpanzees [35,36,59]. However, the direction of hand use preference in macaques depended on species; while left-handedness predominated in rhesus macaques, ambidexterity predominated in cynomolgus macaques [37].

More than two-thirds of capuchins, which belong to the New World monkeys, showed a preference for left-hand use. The right-lateralized cs depth was obtained in left-handed, but not right-handed capuchins [38,39]. Among baboons, which belong to the genus *Papio* in the Old World monkeys, right-handed and left-handed individuals were equally prevalent. Additionally, the cs depth was lateralized contralaterally to the handedness in baboons [40]. On the other hand, the length and/or surface area of the cs were lateralized contralaterally to the handedness in humans [21,22]. Although handedness may be associated with the lateralization of cerebral sulci demarcating the precentral gyrus, the direction of preferred hands and the affected sides of sulci vary depending on the primate species. The sulcal asymmetry in the precentral cortex may shift from rightward to leftward after the split between the Old World monkeys and great apes, which may be relevant to the acquisition of the right-handedness.

The asymmetry of the sf has reportedly been positively associated with the asymmetry of the planum temporale (ptp) [28]. The ptp, cytoarchitecturally homologous to Brodmann's area 22, was left-lateralized in the surface area and gray matter volume, particularly among right-handed chimpanzees and humans [41]. A positive association between the sf and ptp asymmetries was also revealed in chimpanzees via the population-level leftward asymmetry of the sf depth and surface area [29]. In the Old World monkeys, the ptp volume was left-lateralized in cynomolgus macaques [42] and baboons [43], but not in rhesus macaques, bonnet macaques, and vervet monkeys [44]. However, there were no obvious asymmetries in the sf length in cynomolgus macaques [20] nor the sf depth and surface area in the mixed brain samples from rhesus and bonnet macaques [29]. On the other hand, the handedness, which altered the ptp asymmetry in chimpanzees and humans [41,45], was inconsistent among macaques [37]. While the ptp may have started to lateralize leftward during the evolution of Catarrhini (the common ancestor of the Old World monkeys, great apes, and humans), the sf asymmetry may reflect the enhancement of the ptp left-lateralization by acquiring right-handedness after the evolution of Hominoidea.

5. Possible Mediators for Evaluating Sulcal Asymmetry

One significant strategy for evaluating the sulcal asymmetry is the use of the gyrification index (GI). The GI, which was originally designed by Zilles et al. (1988) [46], is the most commonly used quantitative parameter for evaluating the cortical convolution and better understanding the interspecies differences, ontogeny, sexual dimorphism, and pathological statuses of neurodevelopmental and neurological disorders, as well a linkage with the functional development of particular cortical regions [34]. The GI is defined by the ratio of cortical contours including and excluding sulcal grooves on coronal slices of the cerebrum. Generally, the GI is calculated as the mean value of all coronal slices throughout the cerebral hemisphere. This value is referred to as the "global-GI" in this manuscript. The global-GI is close to 1 in lissencephalic mammals such as rats (1.02) and mice (1.01), while the marked highest value is found in the Pacific pilot whale (5.55). Meanwhile, the global-GI is ~2.56 in humans and between 1.78 and 1.81 in macaques [34]. On the other hand, the sulcal-GI is defined by the ratio of the cortical contours including specific primary sulcus to that excluding other sulcal grooves [9]. The sulcal-GI reflects the length, depth, width, and surface areas of specific sulci as well as the local expansion of particular cortical regions. For example, the sulcal-GI is lower when the sulcal width increases, even if no changes are observed in the sulcal length, depth, and surface area [9]. Furthermore, late-emerging sulcal groups infolded in multimodal association cortices showed significant correlations between sulcal-GIs and cortical volumes in cynomolgus macaques [16]. Further studies will be needed to evaluate the sulcal asymmetry using the sulcal-GI to obtain new insights into the brain lateralization of various species of mammals, including primates.

6. Conclusions

The gyrification of the cortical regions was associated with evolutionary expansion [16] and was higher in humans (global-GI = 2.56) and great apes (global-GI = 2.17–2.31) compared to the Old World monkeys (global GI = 1.64–2.27) [34]. Although the global-GI was correlated with larger brain size within the primate order [18,34,55], the increasing gyrification occurred regionally rather than throughout the entire brain. Particularly, the sulcal infolding has progressed in multimodal association cortices [16]. Therefore, the sulcal asymmetry may reflect the species-related specification of the cortical morphology and function, allowed by the evolutionary expansion in higher primates. On the other hand, morphological and functional lateralization was observed in other brain regions. Several studies have reported on the cerebellar asymmetry in mammalian species such as humans [47], chimpanzees [38], capuchins [38], ferrets [60–62], and dogs [63]. Cerebellar asymmetry is known to involve the large-scale fiber connections, which have a contralateral linkage to the cerebellar hemispheres with the cerebral association cortices [64]. An association between the sulcal asymmetry and the lateralization of other brain regions such as the cerebellum will provide new insights into the evolution of the mammalian brain regarding species-related specialization of the brain morphology and function.

Funding: No funding was received.

Conflicts of Interest: There are no conflict of interest.

References

1. Rogers, L.J.; Vallortigara, G.; Andrew, R.J. *Divided Brains-The Biology and Behaviour of Brain Asymmetries*; Cambridge Univ. Press: New York, NY, USA, 2013.
2. Chi, J.G.; Dooling, E.C.; Gilles, F.H. Gyral development of the human brain. *Ann. Neurol.* **1977**, *1*, 86–93. [CrossRef]
3. Pillay, P.; Manger, P.R. Order-specific quantitative patterns of cortical gyrification. *Eur. J. Neurosci.* **2007**, *25*, 2705–2712. [CrossRef] [PubMed]
4. Sawada, K.; Fukunishi, K.; Kashima, M.; Imai, N.; Saito, S.; Sakata-Haga, H.; Aoki, I.; Fukui, Y. Neuroanatomic and MRI references for normal development of cerebral sulci of laboratory primate, cynomolgus monkeys (*Macaca fascicularis*). *Congenit. Anom. (Kyoto)* **2012**, *52*, 16–27. [CrossRef] [PubMed]
5. Sawada, K.; Hikishima, K.; Murayama, A.Y.; Okano, H.J.; Sasaki, E.; Okano, H. Fetal sulcation and gyrification in common marmosets (Callithrix jacchus) obtained by ex vivo magnetic resonance imaging. *Neuroscience* **2014**, *257*, 158–174. [CrossRef] [PubMed]
6. Ferrer, I.; Hernández-Martí, M.; Bernet, E.; Galofré, E. Formation and growth of the cerebral convolutions I. Postnatal development of the median-suprasylvian gyrus and adjoining sulci in the cat. *J. Anat.* **1988**, *160*, 89–100.
7. Wosinski, M.; Schleicher, A.; Zilles, K. Quantitative analysis of gyrification of cerebral cortex in dogs. *Neurobiology* **1996**, *4*, 441–468.
8. Sawada, K.; Watanabe, M. Development of cerebral sulci and gyri in ferrets (*Mustela putorius*). *Congenit. Anom. (Kyoto)* **2012**, *52*, 168–175. [CrossRef]
9. Sawada, K.; Aoki, I. Biphasic aspect of sexually dimorphic ontogenetic trajectory of gyrification in the ferret cerebral cortex. *Neuroscience* **2017**, *364*, 71–81. [CrossRef]
10. Louw, G.J. The development of the sulci and gyri of the bovine cerebral hemispheres. *Anat. Histol. Embryol.* **1989**, *18*, 246–264. [CrossRef]
11. Marino, L.; Connor, R.C.; Fordyce, R.E.; Herman, L.M.; Hof, P.R.; Lefebvre, L.; Lusseau, D.; McCowan, B.; Nimchinsky, E.A.; Pack, A.A.; et al. Cetaceans have complex brains for complex cognition. *PLoS Biol.* **2007**, *5*, e139. [CrossRef]
12. Mortensen, H.S.; Pakkenberg, B.; Dam, M.; Dietz, R.; Sonne, C.; Mikkelsen, B.; Eriksen, N. Quantitative relationships in delphinid neocortex. *Front. Neuroanat.* **2014**, *8*, 132. [CrossRef] [PubMed]
13. Kochounov, P.; Castro, C.; Davis, D.; Dudley, D.; Brewer, J.; Zhang, Y.; Kroenke, C.D.; Purdy, D.; Fox, P.T.; Simerly, C.; et al. Mapping primary gyrogenesis during fetal development in primate brains: High-resolution in utero structural MRI of fetal brain development in pregnant baboons. *Front. Neurosci.* **2010**, *4*, 20.
14. Kochounov, P.; Glahn, D.C.; Fox, P.; Lancaster, J.L.; Saleem, K.; Shelledy, W.; Zilles, K.; Thompson, P.M.; Coulon, O.; Mangin, J.F.; et al. Genetics of primary cerebral gyrification: Heritability of length, depth and area of primary sulci in an extended pedigree of baboons. *Neuroimage* **2010**, *53*, 1126–1134. [CrossRef]

15. Clark, G.M.; Mackay, C.E.; Davidson, M.E.; Iversen, S.D.; Collinson, S.L.; James, A.C.; Roberts, N.; Crow, T.J. Paracingulate sulcus asymmetry; sex difference, correlation with semantic fluency and change over time in adolescent onset psychosis. *Psychiatry Res.* **2010**, *184*, 10–15. [CrossRef] [PubMed]
16. Sawada, K.; Fukunishi, K.; Kashima, M.; Imai, N.; Saito, S.; Aoki, I.; Fukui, Y. Regional difference in sulcal infolding progression correlated with cerebral cortical expansion in cynomolgus monkey fetuses. *Congenit. Anom. (Kyoto)* **2017**, *57*, 114–117. [CrossRef] [PubMed]
17. Rogers, J.; Kochunov, P.; Zilles, K.; Shelledy, W.; Lancaster, J.; Thompson, P.; Duggirala, R.; Blangero, J.; Fox, P.T.; Glahn, D.C. On the genetic architecture of cortical folding and brain volume in primates. *Neuroimage* **2010**, *53*, 1103–1108. [CrossRef] [PubMed]
18. Sawada, K.; Sun, X.Z.; Fukunishi, K.; Kashima, M.; Satio, S.; Sakata-Haga, H.; Tokado, H.; Aoki, I.; Fukui, Y. Ontogenetic pattern of gyrification in fetuses of cynomolgus monkeys. *Neuroscience* **2010**, *167*, 735–740. [CrossRef]
19. Liu, T.; Wen, W.; Zhu, W.; Trollor, J.; Reppermund, S.; Crawford, J.; Jin, J.S.; Luo, S.; Brodaty, H.; Sachdev, P. The effects of age and sex on cortical sulci in the elderly. *Neuroimage* **2010**, *51*, 19–27. [CrossRef]
20. Imai, N.; Sawada, K.; Fukunishi, K.; Sakata-Haga, H.; Fukui, Y. Sexual dimorphism of sulcal length asymmetry in cerebrum of adult cynomolgus monkeys (*Macaca fascicularis*). *Congenit. Anom. (Kyoto)* **2011**, *51*, 161–166. [CrossRef]
21. Foundas, A.L.; Hong, K.; Leonard, C.M.; Heilman, K.M. Hand preference and magnetic resonance imaging asymmetries of the central sulcus. *Neuropsychiatry Neuropsychol. Behav. Neurol.* **1998**, *11*, 65–71.
22. Klöppel, S.; Mangin, J.F.; Vongerichten, A.; Frackowiak, R.S.; Siebner, H.R. Nurture versus nature: Long-term impact of forced right-handedness on structure of pericentral cortex and basal ganglia. *J. Neurosci.* **2010**, *30*, 3271–3275. [CrossRef] [PubMed]
23. Fornito, A.; Wood, S.J.; Whittle, S.; Fuller, J.; Adamson, C.; Saling, M.M.; Velakoulis, D.; Pantelis, C.; Yücel, M. Variability of the paracingulate sulcus and morphometry of the medial frontal cortex: Associations with cortical thickness, surface area, volume, and sulcal depth. *Hum. Brain Mapp.* **2008**, *29*, 222–236. [CrossRef] [PubMed]
24. Amiez, C.; Wilson, C.R.E.; Procyk, E. Variations of cingulate sulcal organization and link with cognitive performance. *Sci. Rep.* **2018**, *8*, 1–13. [CrossRef] [PubMed]
25. Bonte, M.; Frost, M.A.; Rutten, S.; Ley, A.; Formisano, E.; Goebel, R. Development from childhood to adulthood increases morphological and functional inter-individual variability in the right superior temporal cortex. *Neuroimage* **2013**, *83*, 739–750. [CrossRef] [PubMed]
26. Bodin, C.; Takerkart, S.; Belin, P.; Coulon, O. Anatomo-functional correspondence in the superior temporal sulcus. *Brain Struct. Funct.* **2018**, *223*, 221–232. [CrossRef] [PubMed]
27. Corballis, M.C. *The Lopsided Brain: Evolution of the Generative Mind*; Oxford University Press: New York, NY, USA, 1992.
28. Cantalupo, C.; Pilcher, D.L.; Hopkins, W.D. Are planum temporale and sylvian fissure asymmetries directly related? A MRI study in great apes. *Neuropsychologia* **2003**, *41*, 1975–1981. [CrossRef]
29. Bogart, S.L.; Mangin, J.F.; Schapiro, S.J.; Reamer, L.; Bennett, A.J.; Pierre, P.J.; Hopkins, W.D. Cortical sulci asymmetries in chimpanzees and macaques: A new look at an old idea. *Neuroimage* **2012**, *61*, 533–541. [CrossRef]
30. Bonan, I.; Argenti, A.M.; Duyme, M.; Hasboun, D.; Dorion, A.; Marsault, C.; Zouaoui, A. Magnetic resonance imaging of cerebral central sulci: A study of monozygotic twins. *Acta Genet. Med. Gemellol. (Roma)* **1998**, *47*, 89–100. [CrossRef]
31. Kasprian, K.; Langs, G.; Brugger, P.C.; Bittner, M.; Weber, M.; Arantes, M.; Prayer, D. The prenatal origin of hemispheric asymmetry: An in utero neuroimaging study. *Cereb. Cortex* **2011**, *21*, 1076–1083. [CrossRef]
32. Sawada, K.; Fukunishi, K.; Kashima, M.; Saito, S.; Sakata-Haga, H.; Aoki, I.; Fukui, Y. Fetal gyrification in cynomolgus monkeys: A concept of developmental stages of gyrification. *Anat. Rec. (Hoboken)* **2012**, *295*, 1065–1074. [CrossRef]
33. Sakamoto, K.; Sawada, K.; Fukunishi, K.; Imai, N.; Sakata-Haga, H.; Fukui, Y. Postnatal change in sulcal length asymmetry in cerebrum of cynomolgus monkeys (*Macaca fascicularis*). *Anat. Rec. (Hoboken)* **2014**, *297*, 200–207. [CrossRef] [PubMed]
34. Zilles, K.; Palomero-Gallagher, N.; Amunts, K. Development of cortical folding during evolution and ontogeny. *Trends Neurosci.* **2013**, *36*, 275–284. [CrossRef] [PubMed]

35. Lonsdorf, E.V.; Hopkins, W.D. Wild chimpanzees show population level handedness for tool use. *Proc. Natl. Acad. Sci. USA* **2005**, *102*, 12634–12638. [CrossRef] [PubMed]
36. Corp, N.; Byrne, R.W. Sex difference in chimpanzee handedness. *Am. J. Phy. Anthropol.* **2004**, *123*, 62–68. [CrossRef] [PubMed]
37. Chatagny, P.; Badoud, S.; Kaeser, M.; Gindrat, A.D.; Savidan, J.; Fregosi, M.; Moret, V.; Roulin, C.; Schmidlin, E.; Rouiller, E.M. Distinction between hand dominance and hand preference in primates: A behavioral investigation of manual dexterity in nonhuman primates (macaques) and human subjects. *Brain Behav.* **2013**, *3*, 575–595. [CrossRef] [PubMed]
38. Phillips, K.; Hopkins, W.D. Exploring the relationship between cerebellar asymmetry and handedness in chimpanzees (*Pan troglodytes*) and capuchins (*Cebus apella*). *Neuropsychologia* **2007**, *45*, 2333–2339. [CrossRef] [PubMed]
39. Phillips, K.A.; Thompson, C.R. Hand preference for tool-use in capuchin monkeys (*Cebus apella*) is associated with asymmetry of the primary motor cortex. *Am. J. Primatol.* **2013**, *75*, 435–440. [CrossRef]
40. Margiotoudi, K.; Marie, D.; Claidière, N.; Coulon, O.; Roth, M.; Nazarian, B.; Lacoste, R.; Hopkins, W.D.; Molesti, S.; Fresnais, P.; et al. Handedness in monkeys reflects hemispheric specialization within the central sulcus. An in vivo MRI study in right- and left-handed olive baboons. *Cortex* **2019**, *118*, 203–211. [CrossRef]
41. Hopkins, W.D.; Nir, T. Planum temporale surface area and grey matter asymmetries in chimpanzees (*Pan troglodytes*): The effect of handedness and comparison within findings in humans. *Behav. Brain Res.* **2010**, *208*, 436–443. [CrossRef]
42. Gannon, P.J.; Holloway, R.L.; Broadfield, D.C.; Braum, A.R. Asymmetry of chimpanzee planum temporale: Humanlike pattern of Wernicke's brain language area homolog. *Science* **1998**, *279*, 220–222. [CrossRef]
43. Marie, D.; Roth, M.; Lacoste, R.; Nazarian, B.; Bertello, A.; Anton, J.L.; Hopkins, W.D.; Margiotoudi, K.; Love, S.A.; Meguerditchian, A. Left brain asymmetry of the planum temporale in a nonhominid primate: Redefining the origin of brain specialization for language. *Cereb. Cortex* **2018**, *28*, 1808–1815. [CrossRef] [PubMed]
44. Lyn, H.L.; Pierre, P.; Bennett, A.J.; Fears, S.C.; Woods, R.P.; Hopkins, W.D. Planum temporale grey matter asymmetries in chimpanzees (*Pan troglodytes*), vervet (*Chlorocebus aethiops sabaeus*), rhesus (*Macaca mulatta*) and bonnet (*Macaca radiata*) monkeys. *Neuropsychologia* **2011**, *49*, 2004–2012. [CrossRef] [PubMed]
45. Hopkins, W.D. Neuroanatomical asymmetries and handedness in chimpanzees (*Pan troglodytes*): A case for continuity in the evolution of hemispheric specialization. *Ann. N. Y. Acad. Sci.* **2013**, *1288*, 17–35. [CrossRef] [PubMed]
46. Zilles, K.; Armstrong, E.; Schleicher, A.; Kretschmann, H.J. The human pattern of gyrification in the cerebral cortex. *Anat. Embryol.* **1988**, *179*, 173–179. [CrossRef] [PubMed]
47. Rosch, R.E.; Ronan, L.; Cherkas, L.; Gurd, J.M. Cerebellar asymmetry in a pair of monozygotic handedness-discordant twins. *J. Anat.* **2010**, *217*, 38–47. [CrossRef]
48. Kelava, I.; Lewitus, E.; Huttner, W.B. The aecondary loss of gyrencephaly as an example of evolutionary phenotypical reversal. *Front. Neuroanat.* **2013**, *7*, 16. [CrossRef]
49. Yeni-Komshian, G.H.; Benson, D.A. Anatomical study of cerebral asymmetry in the temporal lobe of humans, chimpanzees, and rhesus monkeys. *Science* **1976**, *192*, 387–389. [CrossRef]
50. Falk, D.; Cheverud, J.; Vannier, M.W.; Conroy, G.C. Advanced computer graphics technology reveals cortical asymmetry in endocasts of rhesus monkeys. *Folia Primatol.* **1986**, *4*, 98–103. [CrossRef]
51. Falk, D.; Hildebolt, C.; Cheverud, J.; Vannier, M.; Helmkamp, R.C.; Konigsberg, L. Cortical asymmetries in frontal lobes of rhesus monkeys (*Macaca mulatta*). *Brain Res.* **1990**, *512*, 40–45. [CrossRef]
52. Heilbroner, P.L.; Holloway, R.L. Anatomical brain asymmetries in New World and Old World monkeys: Stages of temporal lobe development in primate evolution. *Am. J. Phys. Anthropol.* **1988**, *76*, 39–48. [CrossRef]
53. Heilbroner, P.L.; Holloway, R.L. Anatomical brain asymmetry in monkeys: Frontal, temporoparietal, and limbic cortex in Macaca. *Am. J. Phys. Anthropol.* **1989**, *80*, 203–211. [CrossRef] [PubMed]
54. Goulas, A.; Stiers, P.; Hutchison, R.M.; Everling, S.; Petrides, M.; Margulies, D.S. Intrinsic functional architecture of the macaque dorsal and ventral lateral frontal cortex. *J. Neurophysiol.* **2017**, *117*, 1084–1099. [CrossRef] [PubMed]
55. Armstrong, E.; Schleicher, A.; Omran, H.; Curtis, M.; Zilles, K. The ontogeny of human gyrification. *Cereb. Cortex* **1995**, *5*, 56–63. [CrossRef] [PubMed]

56. Sholl, S.A.; Kim, K.L. Androgen receptors are differentially distributed between right and left cerebral hemispheres of the fetal male rhesus monkey. *Brain Res.* **1990**, *516*, 122–126. [CrossRef]
57. Hopkins, W.D.; Pilcher, D.L.; MacGregor, L. Sylvian fissure asymmetries in nonhuman primates revisited: A comparative MRI study. *Brain Behav. Evol.* **2000**, *56*, 293–299. [CrossRef]
58. Li, L.; Preuss, T.M.; Rilling, J.K.; Hopkins, W.D.; Glasser, M.F.; Kumar, B.; Nana, R.; Zhang, X.; Hu, X. Chimpanzee (*Pan troglodytes*) precentral corticospinal system asymmetry and handedness: A diffusion magnetic resonance imaging study. *PLoS ONE* **2010**, *21*, e12886. [CrossRef]
59. Humle, T.; Matsuzawa, T. Laterality in hand use across four tool use behaviors among the wild chimpanzees of Bossou, Guinea, West Africa. *Am. J. Primatol.* **2009**, *71*, 40–48. [CrossRef]
60. Sawada, K.; Horiuchi-Hirose, M.; Saito, S.; Aoki, I. Male prevalent enhancement of leftward asymmetric development of the cerebellar cortex in ferrets (*Mustela putorius*). *Laterality* **2015**, *20*, 723–737. [CrossRef]
61. Sawada, K.; Aoki, I. Age-dependent sexually-dimorphic asymmetric development of the ferret cerebellar cortex. *Symmetry* **2017**, *9*, 40. [CrossRef]
62. Sawada, K.; Kamiya, S.; Aoki, I. Asymmetry of cerebellar lobular development in ferrets. *Symmetry* **2020**, *12*, 735. [CrossRef]
63. Koyun, N.; Aydinlioğlu, A.; Aslan, K. A morphometric study on dog cerebellum. *Neurol. Res.* **2011**, *33*, 220–224. [CrossRef] [PubMed]
64. Wang, D.; Buckner, R.L.; Liu, H. Cerebellar asymmetry and its relation to cerebral asymmetry estimated by intrinsic functional connectivity. *J. Neurophysiol.* **2013**, *109*, 46–57. [CrossRef] [PubMed]

Article

The Topology of Pediatric Structural Asymmetries in Language-Related Cortex

Mark A. Eckert [1,*], Federico Iuricich [2], Kenneth I. Vaden Jr. [1], Brittany T. Glaze [1] and Dyslexia Data Consortium

1 Hearing Research Program, Department of Otolaryngology-Head and Neck Surgery, Medical University of South Carolina, Charleston, SC 29425, USA; Vaden@musc.edu (K.I.V.J.); glaze@musc.edu (B.T.G.)

2 Visual Computing Division, School of Computing, Clemson University, Clemson, SC 29634, USA; fiurici@g.clemson.edu

* Correspondence: eckert@musc.edu

Received: 21 September 2020; Accepted: 28 October 2020; Published: 31 October 2020

Abstract: Structural asymmetries in language-related brain regions have long been hypothesized to underlie hemispheric language laterality and variability in language functions. These structural asymmetries have been examined using voxel-level, gross volumetric, and surface area measures of gray matter and white matter. Here we used deformation-based and persistent homology approaches to characterize the three-dimensional topology of brain structure asymmetries within language-related areas that were defined in functional neuroimaging experiments. Persistence diagrams representing the range of values for each spatially unique structural asymmetry were collected within language-related regions of interest across 212 children (mean age (years) = 10.56, range 6.39–16.92; 39% female). These topological data exhibited both leftward and rightward asymmetries within the same language-related regions. Permutation testing demonstrated that age and sex effects were most consistent and pronounced in the superior temporal sulcus, where older children and males had more rightward asymmetries. While, consistent with previous findings, these associations exhibited small effect sizes that were observable because of the relatively large sample. In addition, the density of rightward asymmetry structures in nearly all language-related regions was consistently higher than the density of leftward asymmetric structures. These findings guide the prediction that the topological pattern of structural asymmetries in language-related regions underlies the organization of language.

Keywords: structural asymmetry; language laterality; topological data analysis; persistent homology

---

1. Introduction

Significant advances have been made in our understanding of structural and functional cerebral asymmetries [1,2]. Nonetheless, there remains some mystery about the development of structural asymmetries and their significance for brain organization and individual variation in behavior [3]. Specifically, there is limited understanding about the significance of left cerebral hemisphere dominance for some language functions, whether and how structural asymmetries may underlie this specialization, what functional advantages structural asymmetries may confer, and how intrahemispheric organization influences language functions [4].

A defining feature of the human brain is its asymmetric structure, including in language-related brain regions that exhibit leftward asymmetric activation during language tasks [5–7]. Leftward anterior insula asymmetry and rightward superior temporal sulcus asymmetry, for example, have been consistently observed across surface area and volumetric measurement methods where univariate voxel-level analyses were performed in pediatric [8,9] and adult samples [10–13]. While variation in insular asymmetry has been related to estimates of language dominance and asymmetric activation in at least three studies where volumetric asymmetries were examined [14–16], measures of other language-related regions have been inconsistently associated with language laterality [14,16–20]. One explanation for these inconsistent findings is the influence of demographic and cognitive variables that have been related to structural asymmetries.

Sex has been related to structural asymmetries in multiple voxel-based studies [10,13,21], although non-significant results have been reported [8,11], perhaps because sex differences in brain size may explain locally specific sex effects in some samples [22]. The magnitude of sex differences may also depend on age. Males exhibited more pronounced rightward asymmetry in cortical surface area than females, but this effect was most evident in younger participants [13]. Age has also been related to structural asymmetries, and these effects appear to be dependent on sample size, the range of age, and perhaps the method of analysis. For example, increasing age was modestly associated with more leftward asymmetry in superior temporal gyrus cortical thickness and more rightward asymmetry in superior temporal sulcus surface area [13].

Variation in structural asymmetries has also been related to oral language abilities, although inconsistently. An early imaging study demonstrated higher verbal comprehension (VIQ) when a left parieto-occipital region was larger than the right in a reading disability sample [23]. Leftward posterior superior temporal gyrus asymmetry was also significantly associated with an estimate of VIQ [9]. Better expressive and receptive language has been observed in children with less rightward gray matter volume asymmetry of the pars triangularis (Brodmann area 45), which appeared largely due to the volume of the right pars triangularis [22]. In addition, change towards more rightward pars triangularis cortical thickness, due to reduced left cortical thickness, was associated with larger gains in language skills in children [24]. This smattering of associations with asymmetries across the cortex involved different methods and different sampling approaches. Still, there have been relatively few consistent findings, and some non-significant results have been reported [25].

The structural asymmetry findings described above all involved gross morphometric approaches where asymmetry values were averaged across a relatively large region of the brain, or the analyses involved voxel-level asymmetry comparisons. Structural asymmetries, at least volumetric or surface area asymmetries, have unique spatial distributions and exhibit three-dimensional structure. Topological data analysis is well-suited for characterizing the morphology of brain structure asymmetries. That is, spatially complex structural asymmetries can be measured with topological approaches that distill this complexity, as well as some redundant information at the voxel-level, to a minimal but precise representation of the asymmetry.

Persistent homology, the most widely adopted tool in topological data analysis is a mathematical tool rooted in algebraic topology, which can be used to characterize the persistence of a structure across a given dimension. This includes voxel values in images [26] where a structure has voxel values with a local minima and local maxima. In the case of voxel-based structural asymmetries, a leftward asymmetry in a left hemisphere language-related region would be represented with a range of positive values defining a hyperintense cluster of voxels. Likewise, rightward asymmetries would be represented in the same left hemisphere region with a range of negative values defining a hypointense cluster of voxels or a void. Persistent homology is used to measure the local maxima and minima of the contrast values that define leftward or rightward asymmetries. An example of the persistent homology measure of leftward and rightward asymmetries is presented later in the Methods.

The overarching goal of this descriptive study was to characterize the topology of deformation-based or volumetric asymmetries within language-related areas that exhibit asymmetric patterns of activity. Specifically, analyses were performed to examine the extent to which topological asymmetries were observed in 18 core language-regions, as defined by a functional language laterality atlas [6]. Sex, age, and VIQ associations with the topological asymmetries were examined to determine the extent to which these associations would replicate previous findings [13] and, more generally, to characterize the extent to which these factors explain some asymmetric variation in language-related brain regions.

2. Materials and Methods

Participants. Retrospective data from 212 children (mean age = 10.56 years, range 6.39–16.92; 39% female) across 10 research sites were included in this study. These data were collected as part of a study to develop methods for multi-site retrospective neuroimaging studies and are part of a Dyslexia Data Consortium database (www.dyslexiadata.org). This normative study focused on children with typically developing language skills. Informed consent and institutional approval for the original research was obtained at each institution. The data were de-identified prior to data sharing, which was approved by the contributing institution and the Medical University of South Carolina Institutional Review Board. Data from a subset of these children has been reported previously [27–30].

Imaging Data and Image Processing. The T1-weighted image acquisition parameters for each research site are listed in Table 1. The images were denoised [31], bias field corrected using the Statistical Parametric Mapping (SPM) non-uniformity (bias/apply) functions, and then rigidly aligned to the MNI 152 T1 1mm template using the SPM coregistration function. These images, and their left-right flipped copies, were used to create a study-specific symmetric template using Advanced Normalization Tools (ANTs v 2.0) [32]. The native space images were also segmented using the SPM new segment function so that a total brain volume estimate could be obtained from the sum of the gray and white matter images.

Table 1. T1-weighted image parameters from the 10 study sites.

Site	Manufacturer	Field Strength (T)	Image Dimension (mm)	Slice Thickness (mm)	TR (msec)	TE (msec)	Flip Angle (deg)
1	Siemens	1.5	256 × 256 × 160	1.60	25.00	4.60	30
2	Siemens	3.0	176 × 240 × 256	0.90	2250.00	3.96	9
3	Siemens	3.0	128 × 256 × 256	1.33	6.60	2.90	8
4a	GE	1.5	124 × 256 × 256	1.20	11.10	2.20	25
4b	GE	1.5	124 × 256 × 256	1.40	11.10	2.20	25
5	Siemens	3.0	160 × 256 × 256	1.00	1600.00	3.37	15
6	Philips	1.5	170 × 256 × 256	1.00	8.02	3.69	7
7	GE	1.5	181 × 217 × 181	1.00	6.00	63.00	–
8	Siemens	1.5	160 × 256 × 256	1.00	2000	3.65	8
9	GE	3.0	256 × 256 × 124	1.20	9.00	2.00	15
10	Philips	3.0	256 × 256 × 120	1.10	10.00	6.00	8

A two-step ANTs normalization procedure was performed where an optimal symmetrical template was created without regard for how much each image would be warped or distorted. This template was used in step 2 to more conservatively align the individual images to the optimal template to prevent over-regularization or grossly distorted anatomical features within each participant's image. Specifically, 1) ten normalization iterations were initially performed to create the optimal template where the mean of the images from the preceding step served as the normalization target for the next iteration [normalization parameters—iterations 1–9 (30 × 90 × 20 × 12); iteration 10 (30 × 90 × 30 × 20); across iterations: cross-correlation metric (mm radius = 4.00); SyN (2, 1, 1); Gaussian regularization (4.00)]. The optimal symmetrical template was then 2) used as a target for a series of three normalization iterations that varied in the magnitude of volumetric displacement, as described in [8] [normalization parameters—iteration 1 (50 × 1 × 1): cross-correlation metric (mm radius = 4.00), SyN (1.00), Gaussian regularization (3.00); iteration 2 (1 × 50 × 1): cross-correlation metric (mm radius = 2.00), SyN (0.75), Gaussian regularization (2.00); iteration 3 (1 × 1 × 40): cross-correlation metric (mm radius = 1.00), SyN (0.50), Gaussian regularization (1.00)].

The warping parameters from each of the three normalization iterations were combined to generate warps that align each native space image to the symmetrical template. These warps also characterized how much volumetric displacement was necessary to move a voxel into the template space. This volumetric displacement was represented by the Jacobian determinant, which was obtained using the ANTSJacobian function that included the linear and nonlinear displacements and was log-scaled. Jacobian values representing the warping of relatively large structures had larger values compared to relatively smaller structures. The Jacobian images were then smoothed with an 8 mm full width at half maximum Gaussian kernel for consistency with previous voxel-based studies [21,33]. Because the images were log-scaled, each Jacobian image could be subtracted by its left-right flipped copy to create a standard asymmetry image where hemispheric differences are scaled by the mean volumetric displacement of the left and right hemispheres. An overview of these image processing procedures is presented in Supplementary Figure S1.

The Jacobian asymmetry results presented later can be interpreted in the context of other voxel-based studies of gray and white matter asymmetries where Jacobian modulation of the gray or white matter voxel values was performed. That is, the voxel-level asymmetries with this Jacobian or deformation-based approach are very similar to the voxel-level asymmetries involving specific tissue types [8]. Given the similar pattern of results between voxel-based gray matter volume asymmetries and surface area asymmetries [9,12], the current volumetric deformation approach was also expected to yield results that are similar to surface area asymmetry results.

The imaging parameters in Table 1 were obtained from de-identified DICOM information when available, otherwise from related manuscripts or image header information, which is why flip angle information is unknown for one site. The relatively long repetition times (TRs) above were used for inversion recovery acquisitions (inversion times: site 2 = 900 msec; site 5 = 640 msec). Site 4a,b data were obtained for 2 different studies from 1 research site. GE—general electric; TR—repetition time; TE—echo time.

Regions of Interest. The [SENSAAS; [6]; https://www.gin.cnrs.fr/en/tools/sensaas/] was used to define language-related ROI. The SENSAAS is based on an atlas of homologous interhemispheric regions that exhibited common patterns of resting state activity [34]. This includes 18 core regions that exhibited leftward activation asymmetries, which appeared to be essential for sentence processing and where lesions can produce language deficits [6]. The SENSAAS ROI are in MNI coordinate space, so the 1 mm T1 MNI template was ANTs normalized into the space of the optimal symmetrical template. The warping parameters were applied to each of the SENSAAS ROI to put them into the same coordinate space as the asymmetry images [Normalization parameters- 1 iteration (30 × 90 × 30 × 20); Gaussian regularization (4.00); SyN (2,1,1)]. To determine the influence of ROI size on the results from the persistent homology measures, the volume of each ROI was determined by summing the voxels in each binary ROI. In addition, the mean asymmetry within each ROI was obtained using MarsBar [35] for comparison to the topological asymmetry data.

Persistent Homology. Persistent homology is a widely used tool, grounded in algebraic topology, to analyze shape features by means of a filtration. Please see [36] for the original description of this approach, which we formalize as follows. Let Γ represent a cubical complex (i.e., the image grid), the filtration is defined as a sequence $\Gamma\hat{}f = \{\Gamma\hat{}i \mid 0 \leq I \leq r\}$, such that $\varnothing = \Gamma\hat{}0 \subseteq \Gamma\hat{}1 \subseteq \cdots \subseteq \Gamma\hat{}r = \Gamma$. When working with images, a filtration is naturally defined as the sequence of sublevel sets of Γ. Intuitively, a filtration corresponds to a growing object obtained by increasing an intensity threshold to points in a 3D image (i.e., voxels), as shown in Figure 1. Voxel intensity is color-coded according to a diverging (blue-red) color map. Low-intensity voxels are introduced first in the filtration, generating seven components (blue regions in Figure 1A). As voxels with higher intensities are added, the seven components merge into one (Figure 1B). Eventually, all voxels on the boundary of the image are introduced, thus creating a cavity (Figure 1C). The border of this cavity is depicted in gray. While shrinking, the cavity splits into two in Figure 1D. The newly born cavity is filled up in Figure 1E. Finally, the last cavity disappears when voxels across intensity values are introduced by the filtration (Figure 1F).

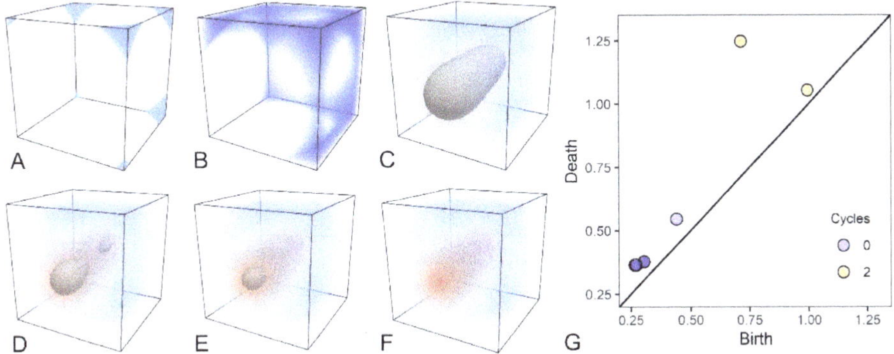

Figure 1. Example of the sublevel set filtration defined on a 3D image and persistence diagram. The intensity of voxels is color-coded according to a diverging (blue-red) color map. Gray surfaces indicate cavities appearing and disappearing in the filtration from (**A–F**). (**G**) The 0–cycle and 2–cycle structures identified in the filtration are presented in the persistence diagram where the x and y axes indicate the contrast values corresponding to the birth and death of each structure. Note the short lifespan of the seven 0–cycle structures with low contrast values in A are clustered at the bottom of the identity line in (**G**). Also note the two 2–cycle structures with higher contrast values are shown in (**G**) with one 2–cycle structure having a longer lifespan (range of contrast values) based on its distance from the identity line.

Persistent homology keeps track of all components, or 0–cycles, and hyper-intense objects, also called 2–cycles, which appear and disappear through filtration. Each cycle is characterized by a pair of indices (i,j), or a persistence pair, that describes when the cycle was generated (birth) and destroyed (death), during filtration. It can be shown that the birth of a 0–cycle corresponds to a local minimum in the image [37]. In contrast, the death of a 2–cycle corresponds to a local maximum. In this asymmetry study where ROI were placed in the left hemisphere, the continuum of asymmetries reflected contralateral to ipsilateral asymmetries. 0–cycles corresponded to contralateral asymmetries and 2–cycles correspond to ipsilateral asymmetries. The lifespan of each cycle defines the range of asymmetry values that represents the asymmetric structure.

The Topology Toolkit [v.0.9.8; [38]] was used to compute persistence pairs from each participant's asymmetry images. Again, 0–cycles characterized voids or contralateral asymmetries, and 2–cycles characterized hyperintense objects or ipsilateral asymmetries in the left hemisphere. For each ROI, we collected all contained persistence pairs. Structures defined by persistence pairs can extend beyond the space of an ROI. For this reason, persistence pairs corresponding to rightward asymmetry (0–cycle) were included in a ROI only if the birth of the structure occurred within the ROI and had a negative birth value to ensure that the structure was rightward asymmetric. Similarly, persistence pairs corresponding to leftward asymmetry (2–cycle) were included only if the peak voxel value or death of the structure occurred within the ROI.

Persistence pairs are commonly visualized using a persistence diagram or scatterplot graph representing each pair as a dot with coordinates (i,j), as shown in Figure 1G. The longer the lifespan of a cycle, the more distant the corresponding point will be from the diagonal in the persistence diagram. Figure 2A shows a child's asymmetry image where there were two leftward asymmetric structures (blue circles) within a superior temporal sulcus ROI. Figure 2B shows another child's asymmetry image where there was a single rightward asymmetric structure in the superior temporal sulcus and no evidence of a leftward asymmetry.

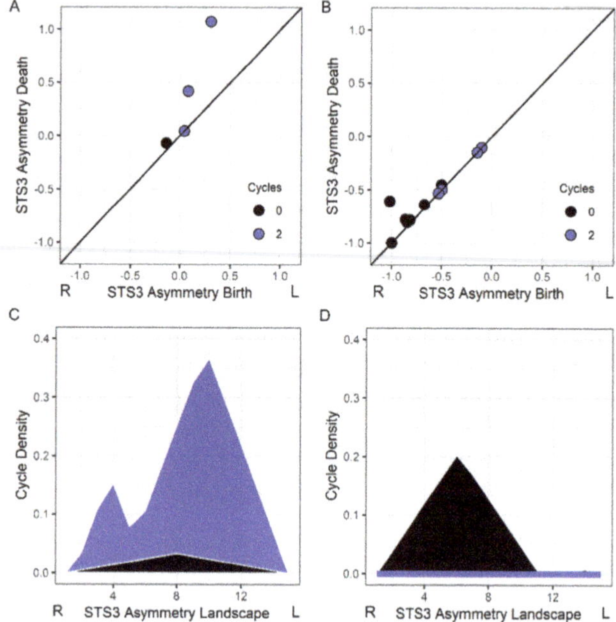

Figure 2. Examples of the persistent homology measures and participant variation for a superior temporal sulcus ROI (STS3). (**A**,**B**) Persistence diagrams of the 0–cycle and 2–cycle persistence pairs, or rightward and leftward asymmetries, for 13.5 year-old male and 11.6 year-old female, respectively. Each circle on the diagonal is typically a very small structure or noise. (**C**,**D**) The persistence pairs from each participant are represented by persistence landscapes (black: 0–cycle; blue: 2–cycle). These landscape data were used to examine varied asymmetries in the ROIs across participants. The R and L indicate rightward and leftward asymmetry, respectively.

Persistence landscapes were used to transform the persistence diagrams into a normalized vector space that represents the lifespan (i.e., magnitude of each asymmetry) and density of asymmetries or cycles across intensity values [39]. The transformation from persistence diagram to landscape allows for statistical comparisons across participants who can have a different number of persistence pairs. The R TDA library (v 1.6.5) [40] was used to create persistence landscapes that represented persistence pairs across 15 positions representing the range of asymmetry values in the images. We later refer to statistical effects at a landscape position because the landscape range represents rightward to leftward asymmetries. Figure 2C,D shows the landscapes for the persistence pairs in Figure 2A,B.

The persistence landscape data was not redundant with the mean asymmetry of all voxels within an ROI, because the persistence homology approach distinguishes leftward and rightward asymmetries that would otherwise be averaged across an ROI. In addition, differences could occur because the approach used here excluded asymmetric structures that were not born in an ROI (rightward asymmetries) or that did not die within an ROI (leftward asymmetries). Thus, the mean asymmetry data exhibited varying degrees of association with the landscape data depending on how many rightward and leftward asymmetric structures were within an ROI. For that reason, we examined the relative association strengths between the persistence landscape and mean asymmetries with age, sex, and VIQ, as described next in the Statistics section.

Statistics. R (v 3.6.0) was used for all statistical analyses. This included the use of the mice library (v 3.6.0) to deal with the 18% multi-site missingness for the VIQ measure [28,41]. Predictor variables in the imputation model were used to create 10 imputed datasets and included age, sex, research site, and total gray matter and white matter volume. The following behavioral variables were not a focus of this asymmetry study but were also included in the multiple imputation model: (1) real word identification (Letter–Word Identification; mean = 109.13, sd = 12.21); (2) pseudoword decoding (Word Attack; mean = 107.99, sd = 11.00), and (3) reading comprehension (Passage Comprehension cloze task; mean = 105.92, sd = 9.40) subtests from the Woodcock–Johnson IIIR [42] and Woodcock Reading Mastery Tests [43]; and (4) rapid naming (mean = 99.67, sd = 11.57) from the Comprehensive Test of Phonological Processing Rapid Automatized Naming or the Rapid Alternating Stimulus Tests [43–45]. The VIQ measures from each of the 10 imputed datasets were highly correlated. For example, VIQ from imputed dataset 1 (mean = 113.00, sd = 14.09) explained 99% of the variance in the VIQ from imputed dataset 10 (mean = 113.06, sd = 14.11). Thus, a single imputation was sufficient and for this reason, VIQ results described later are reported for only the first imputation.

Nonparametric linear regressions were performed to examine age, sex, and VIQ associations with the persistence landscape data. These analyses included the research site as a control variable for potential differences in image acquisition across sites. Follow-up analyses were performed to determine the extent to which total brain volume could explain sex differences. To limit false positive findings and analyze non-normally distributed landscape data, the lmPerm library (v 2.1.0) [46] was used to evaluate the significance of each predictor variable using the lmp function and the permutation F-test probability method. Significant associations were also examined to determine the extent to which the mean asymmetry within each ROI exhibited the same relationship with age, sex, and VIQ as the persistence data. Here we adapted the test of two independent correlation coefficients [47]. The t-score results from each lmp regression result were converted to r values and we used the degrees of freedom from the regression in calculating the significance of differences between landscape asymmetry and mean ROI asymmetry effects.

We also determined the extent to which there was a difference in the density of 0–cycle and 2–cycle asymmetries within each of the 18 language-related ROI. Each of the 0–cycle and 2–cycle landscapes for an ROI were subtracted, and then the mean of this landscape difference was obtained to produce a metric of the relative prevalence of leftward versus rightward asymmetries in each ROI. A one-sample t-test was then performed to examine the extent to which the differences were significant using Wilcoxon Rank non-parametric permutation testing tests with the exactRankTests library (v 0.8-31) [48]. Finally, standard voxel-based morphometry was performed using the CAT12 Toolbox implementation of threshold-free cluster enhancement (TFCE) non-parametric testing [49,50] with a corrected probability threshold of $p < 0.05$ to demonstrate where there were significant leftward and rightward asymmetries at the voxel-level for comparison to the topological results.

Data Availability. Data can be made available with institutional data sharing approvals. Please contact the corresponding author for details, as well as for access to the topological data analysis code used in this study.

3. Results

Figure 3 demonstrates that 0–cycles and 2–cycles, or rightward and leftward asymmetric structures, respectively, were identified in the language-related ROI. The density of asymmetries in each ROI depended on its size. For example, high correlations were observed between ROI volume and the mean density of 0–cycles and 2–cycles at the seventh and twelfth positions along the landscapes (0–cycle landscape position seven: r = 0.884, $p = 1.165 \times 10^{-6}$; 2–cycle landscape position 12: r = 0.839, $p = 1.334 \times 10^{-5}$). However, Figure 3 also shows that there was substantial variation across participants. That is, the size of the ROI determined the density of persistence pairs but not how much the density varied across participants within an ROI.

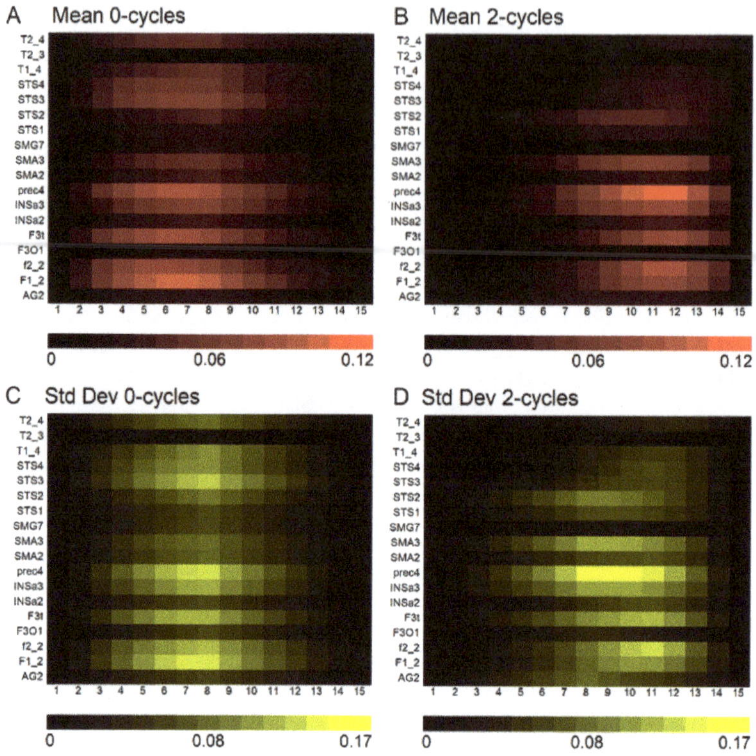

Figure 3. Rightward and leftward asymmetries in the same language-related brain region. (**A**,**B**) Mean 0–cycle or rightward asymmetries and mean 2–cycle or leftward asymmetries. The brighter red regions show where asymmetries were most pronounced or dense across the landscape (x-axis). (**C**,**D**) Std Dev are shown for each of the ROI across the landscape. Acronyms: Angular gyrus (AG2), medial superior frontal gyrus (F1_2), inferior frontal sulcus (f2_2), pars opercularis (F3O1), pars triangularis (F3t), anterior insula (INSa2), anterior insula (INSa3), precentral sulcus (prec4), pre-superior motor areas (SMA2), pre-superior motor areas (SMA3), supramarginal gyrus (SMG7), anterior superior temporal sulcus (STS1; temporal pole), anterior superior temporal sulcus (STS2; anterior to Heschl's gyrus), posterior superior temporal sulcus (STS3; posterior to Heschl's gyrus), posterior superior temporal sulcus (STS4; posterior to Sylvian Fissure), superior temporal gyrus (T1_4), middle temporal gyrus (T2_3), posterior middle temporal gyrus (T2_4).

The 0–cycle and 2–cycle density variances were explained in part by sex, age, and VIQ with small effect sizes. Results from the non-parametric linear regressions across language-related ROI are presented in Appendix A Supplementary Figure S2. This figure shows that some ROI exhibited consistently significant correlations with age and sex across landscape positions. Two of the largest and most spatially contiguous effects across a landscape are shown in Figure 4, where rightward asymmetries (0–cycles) in the posterior and anterior regions of the superior temporal sulcus were more pronounced in males and older children, respectively. These effects were not substantively affected when total brain volume was included in the regression models (e.g., sex and the posterior STS3 (landscape position 6): $t = -3.071$, $p = 0.002$, Cohen's d = 0.44; total brain volume covaried: $t = -2.719$, $p = 0.007$, Cohen's d = 0.39; e.g., age and the anterior STS2 (landscape position 5): $t = 2.479$, $p = 0.014$, Cohen's d = 0.35; total brain volume covaried: $t = 2.494$, $p = 0.013$, Cohen's d = 0.35)). VIQ associations with the asymmetry data included the posterior superior temporal sulcus where increased 0–cycle density was associated with higher VIQ scores (Supplementary Figure S2). However, VIQ associations across ROI were spatially sporadic and small in effect size.

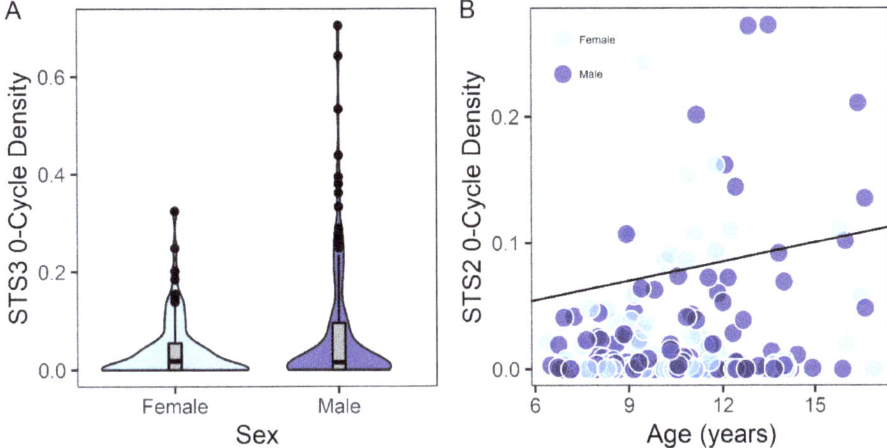

Figure 4. Rightward asymmetry structures (0–cycles) in the posterior and anterior superior temporal sulcus were more pronounced in males and with increased age, respectively. Permutation testing demonstrated that the sex effect in (**A**) and age effect in (**B**) were significant despite the appearance of a subset of cases contributing to the effects. y-axes: STS3 0–cycle Density landscape position 6; STS2 0–cycle density landscape position 5. Acronyms: Anterior superior temporal sulcus (STS2; anterior to Heschl's gyrus), posterior superior temporal sulcus (STS3; posterior to Heschl's gyrus).

The effect sizes for the landscape associations with sex, age, and VIQ were consistently larger than associations observed when using the mean asymmetry within the superior temporal sulcus ROI (sex and STS3: position 6, Cohen's d = 0.44; ROI mean asymmetry: Cohen's d = 0.34; age and STS2: position 5, Cohen's d = 0.35; mean ROI asymmetry, Cohen's d = 0.06). However, the differences in effect sizes between methods were not significantly different (sex: $z = 0.485$, $p = 0.628$; age: $z = 1.444$, $p = 0.149$).

Finally, there were significant differences in the density of 0–cycle compared to 2–cycle asymmetries within each language-related ROI. Figure 5 shows that most participants had a significantly higher density of 0–cycles than 2–cycles across 11 of the 18 ROI and there were no ROI exhibiting a significantly higher density of 2–cycles than 0–cycles. Figure 5 also shows standard voxel-based asymmetry results to provide additional support for the rightward asymmetry topology findings. The 0–cycle > 2–cycle effects appeared to be consistent across age, sex, and VIQ as there were no significant relationships between these variables and the 0–cycle and 2–cycle difference metrics. That is, across children in this sample, there was a preponderance of rightward asymmetric structures within the SENSAAS defined cortical regions that exhibit leftward activity asymmetries during language tasks and this difference was not related to demographic or VIQ variables.

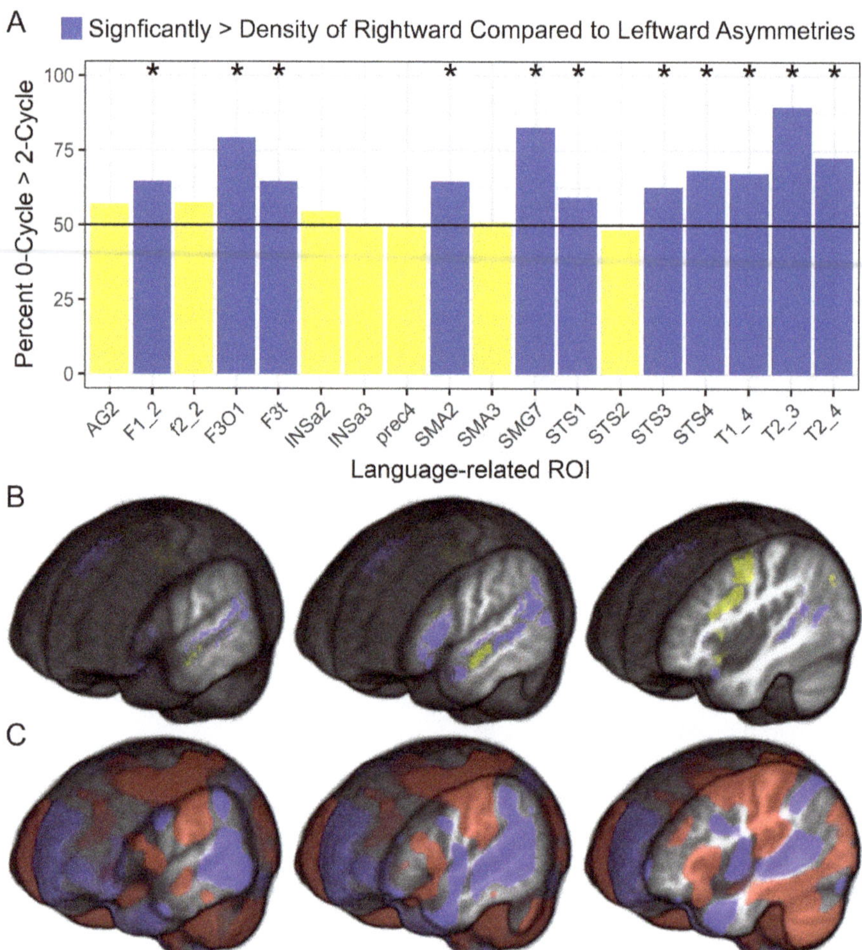

Figure 5. Topological and voxel-based asymmetries. (**A**) There was a consistently greater percentage of children with more rightward asymmetric structures (0–cycles) than leftward asymmetric structures (2–cycles) across language-related ROI. The y-axis is the percentage of children whose summed difference in 0–cycle and 2–cycle landscapes was greater than 0. * $p = 0.05$ to $p < 0.000001$. (**B**) This pattern of effects is presented for each of the ROI on the template image (blue: significantly rightward; yellow: non-significant). (**C**) The topological results largely overlap the whole brain voxel-based asymmetry results (TFCE FWE $p < 0.05$; blue: Rightward asymmetries; red: Leftward asymmetries). Acronyms: Angular gyrus (AG2), medial superior frontal gyrus (F1_2), inferior frontal sulcus (f2_2), pars opercularis (F3O1), pars triangularis (F3t), anterior insula (INSa2), anterior insula (INSa3), precentral sulcus (prec4), pre-superior motor areas (SMA2), pre-superior motor areas (SMA3), supramarginal gyrus (SMG7), anterior superior temporal sulcus (STS1; temporal pole), anterior superior temporal sulcus (STS2; anterior to Heschl's gyrus), posterior superior temporal sulcus (STS3; posterior to Heschl's gyrus), posterior superior temporal sulcus (STS4; posterior to Sylvian Fissure), superior temporal gyrus (T1_4), middle temporal gyrus (T2_3), posterior middle temporal gyrus (T2_4).

4. Discussion

The goal of this study was to characterize the three-dimensional topological asymmetries of language-related regions in a normative pediatric sample. Cortical regions that exhibit consistent patterns of leftward asymmetric activity in language studies exhibited both leftward and rightward structural asymmetries. These topological asymmetries were modestly influenced by demographic factors, particularly rightward superior temporal sulcus asymmetries that were more pronounced in males and older children, even after accounting for total brain volume. Across language-related ROI, there was a consistently higher density of rightward asymmetric structures than leftward asymmetric structures.

The current study differs from previous voxel-based asymmetry studies in an important way. Previous univariate asymmetry studies were designed to examine asymmetries within each brain voxel using spatially-dependent comparisons. The current study used a pure mathematics approach to characterize volumetric asymmetries that could span multiple voxels. Topological approaches are ideally suited for characterizing three-dimensional hyper-intensities and hypo-intensities. Here those structures represented hemispheric differences in the amount of volumetric displacement needed to warp the images to a symmetrical template. At the voxel level, these asymmetry data exhibit the same type of gray matter and white matter asymmetries that have been described previously [8] and appear to yield similar results to surface area asymmetries [12].

Figure 5 shows that the topological data exhibited a similar pattern of results to the voxel-based data, and the few spatially different results between approaches can be explained by the method of limiting the topological analyses to rightward asymmetries born within the ROI and leftward asymmetries that died within the ROI. That is, large deformation differences between hemispheres spanning multiple regions could have a death or peak asymmetry in a region outside of an ROI, which would not be included in the persistence diagram (e.g., large asymmetric deformations of primary somatosensory and motor cortex that extended into the language-related pre-central sulcus in Figure 5C). An advantage of this approach included a significant reduction in the number of variables compared to the voxel-based data. The topological approach also maintained the explicit relationship between the structural asymmetries and their range of voxel values in the persistence diagrams, which demonstrated considerable variation across participants and appeared to have greater sensitivity to individual differences than the mean asymmetry of voxels across an ROI.

Demographic Influences on Structural Asymmetries. The sex effects observed in this study are consistent with findings from voxel-based gray matter volume and surface area studies where males have exhibited more rightward asymmetry in the superior temporal sulcus than females [13,51]. The superior temporal sulcus typically exhibits a rightward asymmetry in voxel-based gray matter volume and deformation studies [8–12]. Here, the sex effect appeared to be driven by the greater prevalence of rightward asymmetries in males rather than more leftward asymmetries in females.

The age effects observed in this study also appear to be consistent with previous findings. For example, older age was associated with less leftward surface area asymmetry in the superior temporal sulcus [13]. In the current study, older children exhibited more rightward-asymmetric structures in the anterior superior temporal sulcus. This result in the current study was driven by a relatively small subset of participants (Figure 4). It seems likely that a large sample size and broad age range is necessary to observe this age effect. Together with the sex effects, demographic factors appeared to have a relatively limited association with topological asymmetries across the language-related regions.

Verbal Comprehension. VIQ associations with the topological asymmetries in language-related regions were relatively small and inconsistently observed across persistence landscapes (Supplementary Figure S2). These results are consistent with non-significant associations reported in studies of white matter diffusion and gray matter density asymmetries [25,52]. One explanation for these results is the limited range of VIQ in this normative sample compared to other studies where measures of VIQ were associated with structural asymmetries (e.g., [9]).

Limitations. Functional asymmetry data were not available for this study. Thus, it was not possible to determine the extent to which the sample exhibited rightward asymmetries in language function that might map to their rightward structural asymmetries. However, given the relatively large sample size of children with relatively average to above average VIQ, it seems likely that the sample is representative of the normal population exhibiting leftward language laterality in brain activity. A related limitation is the absence of a quantitative handedness measure in this study. Variation in hand dexterity and preference may have contributed to variation in the topological asymmetry measures, and there is some evidence from a meta-analysis for a handedness effect on structural asymmetries [13]. However, that same study also demonstrated no effect of handedness on cortical thickness asymmetries in a very large sample size. Finally, we also were unable to determine the extent to which variation in topological asymmetries was influenced by multi-lingual children [53], which, along with other skills such as musicianship, seem important areas of future topological study.

5. Conclusions

Cerebral asymmetries are multi-dimensional and exhibit significant variability across humans [54], including increased variation within local specific regions that can be uncoupled to variation in other regions in comparison to non-human apes [55]. Here we demonstrated rightward and leftward asymmetric structures within locally specific cortical regions that typically exhibit leftward asymmetric activity during language tasks. These structural asymmetries exhibited patterns of variation with sex and age that were largely consistent with previous voxel-based studies, particularly studies where modulated gray matter (i.e., gray matter volume) and surface area asymmetries were examined. The most unexpected result from this study was the greater density of rightward compared to leftward asymmetric structures in language-related regions. In addition to the anterior insula that exhibits leftward asymmetry across structural studies and appears to explain some of the variance in lateralized language expression [14–16], perhaps it is the combination of structural asymmetries that underlies asymmetrical patterns of activity for language expression and reception.

Supplementary Materials: The following are available online at http://www.mdpi.com/2073-8994/12/11/1809/s1, Figures S1 and S2.

Author Contributions: Conceptualization, M.A.E. and F.I.; Methodology, M.A.E and F.I.; Formal Analysis, M.A.E and F.I.; Investigation, M.A.E, F.I., B.T.G.; Resources, Dyslexia Data Consortium; Data Curation, M.A.E and K.I.V.J.; Writing—Original Draft Preparation, M.A.E; Writing—Review and Editing, M.A.E, B.T.G., F.I., K.I.V.J.; Visualization, M.A.E, F.I., K.I.V.J.; Supervision, M.A.E.; Project Administration, M.A.E.; Funding Acquisition, M.A.E. All authors have read and agreed to the published version of the manuscript.

Funding: This work was supported (in part) by the National Institutes of Health (NIH)/Eunice Kennedy Shriver National Institute of Child Health and Human Development (R01 HD 069374) and was conducted in a facility constructed with support from Research Facilities Improvement Program (C06 RR 014516) from the NIH/National Center for Research Resources.

Acknowledgments: Please see www.dyslexiadata.org for more information about and contributors to the Dyslexia Data Consortium who provided the data for this study.

Conflicts of Interest: The authors declare no conflict of interest.

Appendix A

1. Rigidly align T1 image to MNI template

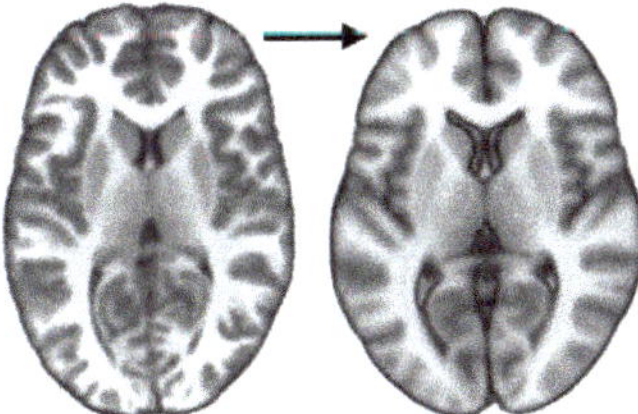

2. Create left-right flipped image

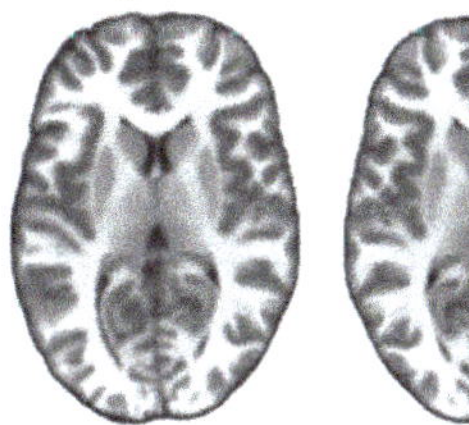

3. Create symmetrical template with all original and flipped images

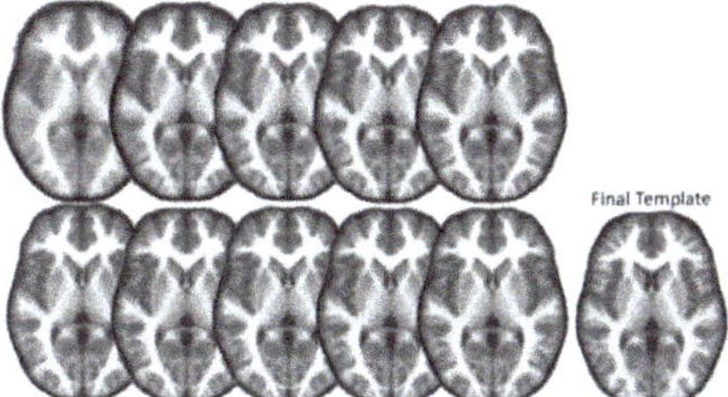

4. Warp each image to the template with 3 successively fine-grained normalization steps

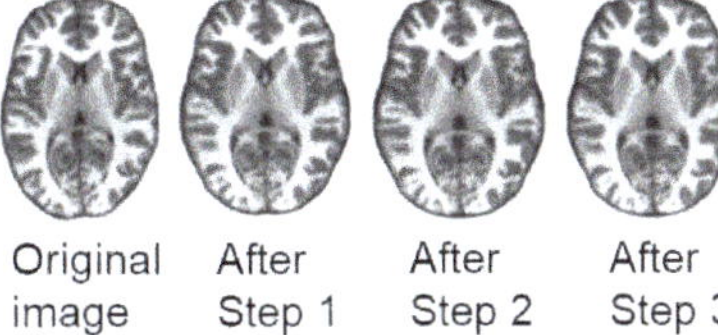

5. Create Jacobian image from normalization warps and smooth

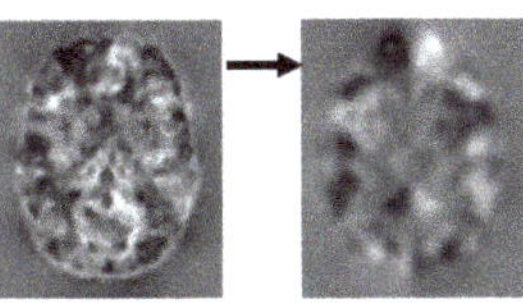

6. After left-right subtracting smoothed images, select data from within each ROI and collect persistence diagrams and landscapes for statistical analysis

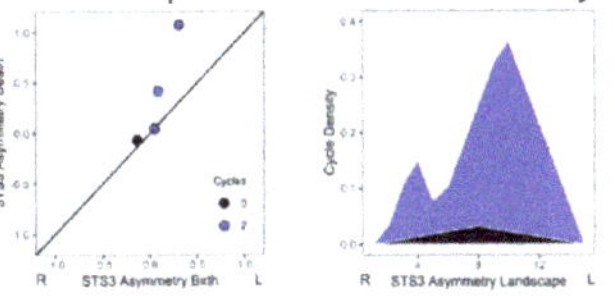

Figure A1. Overview of the image processing steps used to generate structural asymmetry images based on voxel-level volumetric displacement data from warping each T1-weighted image to a study-specific template.

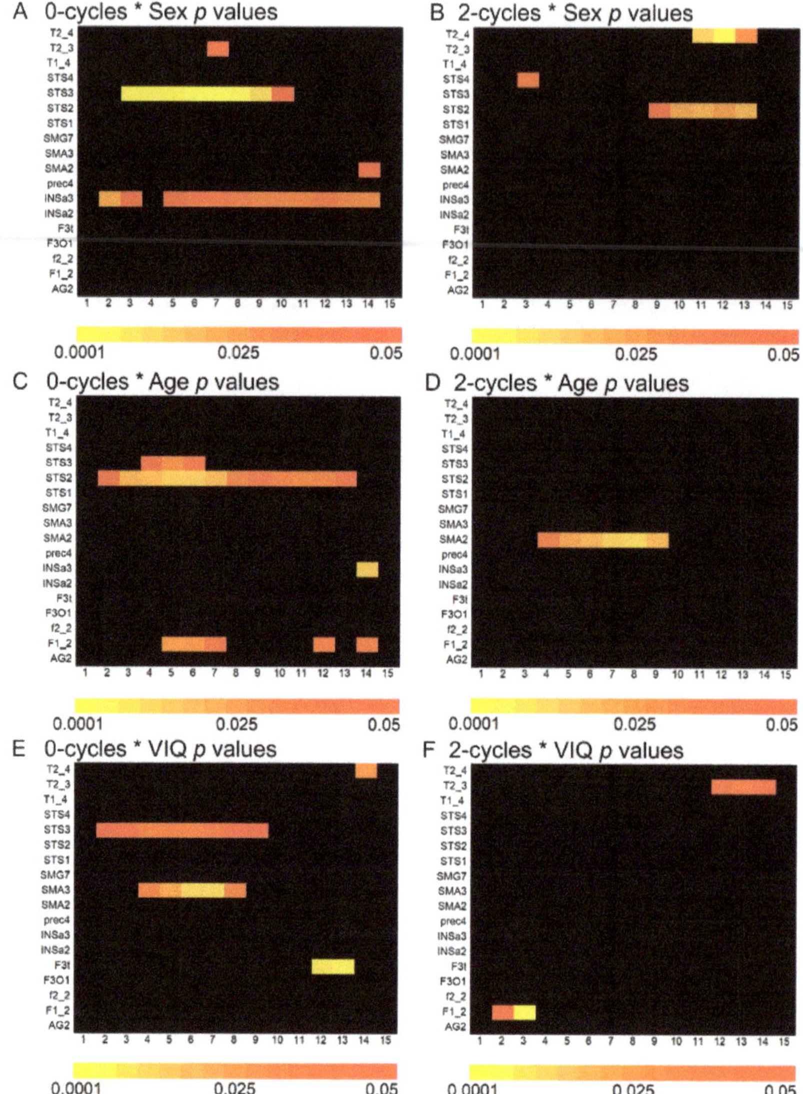

Figure A2. The spatial distribution of landscape associations with sex, age, and VIQ shown in for 0–cycles (**A**,**C**,**E**) and 2–cycles (**B**,**D**,**F**), respectively. Most notable was the greater density of rightward asymmetries in the posterior superior temporal sulcus (STS3) in males compared to females as shown in (**A**) as well as the greater density of rightward asymmetries in the anterior superior temporal sulcus older children (STS2) as shown in (**C**). These non-parametric linear regression results, which included sex, age, VIQ, and research site in the analyses, are color coded for associations exhibiting $p < 0.05$ associations. Acronyms: Angular gyrus (AG2), medial superior frontal gyrus (F1_2), inferior frontal sulcus (f2_2), pars opercularis (F3O1), pars triangularis (F3t), anterior insula (INSa2), anterior insula (INSa3), precentral sulcus (prec4), pre-superior motor areas (SMA2), pre-superior motor areas (SMA3), supramarginal gyrus (SMG7), anterior superior temporal sulcus (STS1; temporal pole), anterior superior temporal sulcus (STS2; anterior to Heschl's gyrus), posterior superior temporal sulcus (STS3; posterior to Heschl's gyrus), posterior superior temporal sulcus (STS4; posterior to Sylvian Fissure), superior temporal gyrus (T1_4), middle temporal gyrus (T2_3), posterior middle temporal gyrus (T2_4).

References

1. Gunturkun, O.; Ocklenburg, S. Ontogenesis of lateralization. *Neuron* **2017**, *94*, 249–263. [CrossRef] [PubMed]
2. Esteves, M.; Lopes, S.S.; Almeida, A.; Sousa, N.; Leite-Almeida, H. Unmasking the relevance of hemispheric asymmetries-Break on through (to the other side). *Prog. Neurobiol.* **2020**, *192*, 101823. [CrossRef] [PubMed]
3. Vingerhoets, G. Phenotypes in hemispheric functional segregation? Perspectives and challenges. *Phys. Life Rev.* **2019**, *30*, 1–18. [CrossRef] [PubMed]
4. Tzourio-Mazoyer, N.; Perrone-Bertolotti, M.; Jobard, G.; Mazoyer, B.; Baciu, M. Multi-factorial modulation of hemispheric specialization and plasticity for language in healthy and pathological conditions: A review. *Cortex* **2017**, *86*, 314–339. [CrossRef] [PubMed]
5. Price, C.J. A review and synthesis of the first 20 years of PET and fMRI studies of heard speech, spoken language and reading. *Neuroimage* **2012**, *62*, 816–847. [CrossRef] [PubMed]
6. Labache, L.; Joliot, M.; Saracco, J.; Jobard, G.; Hesling, I.; Zago, L.; Mellet, E.; Petit, L.; Crivello, F.; Mazoyer, B.; et al. A SENtence Supramodal Areas AtlaS (SENSAAS) based on multiple task-induced activation mapping and graph analysis of intrinsic connectivity in 144 healthy right-handers. *Brain Struct. Funct.* **2019**, *224*, 859–882. [CrossRef] [PubMed]
7. Mazoyer, B.; Zago, L.; Jobard, G.; Crivello, F.; Joliot, M.; Perchey, G.; Mellet, E.; Petit, L.; Tzourio-Mazoyer, N. Gaussian mixture modeling of hemispheric lateralization for language in a large sample of healthy individuals balanced for handedness. *PLoS ONE* **2014**, *9*, e101165. [CrossRef] [PubMed]
8. Eckert, M.A.; Vaden, K.I., Jr.; Dyslexia Data, C. A deformation-based approach for characterizing brain asymmetries at different spatial scales of resolution. *J. Neurosci. Methods* **2019**, *322*, 1–9. [CrossRef]
9. Eckert, M.A.; Lombardino, L.J.; Walczak, A.R.; Bonihla, L.; Leonard, C.M.; Binder, J.R. Manual and automated measures of superior temporal gyrus asymmetry: Concordant structural predictors of verbal ability in children. *Neuroimage* **2008**, *41*, 813–822. [CrossRef]
10. Good, C.D.; Johnsrude, I.; Ashburner, J.; Henson, R.N.; Friston, K.J.; Frackowiak, R.S. Cerebral asymmetry and the effects of sex and handedness on brain structure: A voxel-based morphometric analysis of 465 normal adult human brains. *Neuroimage* **2001**, *14*, 685–700. [CrossRef]
11. Watkins, K.E.; Paus, T.; Lerch, J.P.; Zijdenbos, A.; Collins, D.L.; Neelin, P.; Taylor, J.; Worsley, K.J.; Evans, A.C. Structural asymmetries in the human brain: A voxel-based statistical analysis of 142 MRI scans. *Cereb. Cortex* **2001**, *11*, 868–877. [CrossRef] [PubMed]
12. Maingault, S.; Tzourio-Mazoyer, N.; Mazoyer, B.; Crivello, F. Regional correlations between cortical thickness and surface area asymmetries: A surface-based morphometry study of 250 adults. *Neuropsychologia* **2016**, *93*, 350–364. [CrossRef] [PubMed]
13. Kong, X.Z.; Mathias, S.R.; Guadalupe, T.; Group, E.L.W.; Glahn, D.C.; Franke, B.; Crivello, F.; Tzourio-Mazoyer, N.; Fisher, S.E.; Thompson, P.M.; et al. Mapping cortical brain asymmetry in 17,141 healthy individuals worldwide via the ENIGMA Consortium. *Proc. Natl. Acad. Sci. USA* **2018**, *115*, E5154–E5163. [CrossRef] [PubMed]
14. Keller, S.S.; Roberts, N.; Garcia-Finana, M.; Mohammadi, S.; Ringelstein, E.B.; Knecht, S.; Deppe, M. Can the language-dominant hemisphere be predicted by brain anatomy? *J. Cogn. Neurosci.* **2011**, *23*, 2013–2029. [CrossRef]
15. Keller, S.S.; Roberts, N.; Baker, G.; Sluming, V.; Cezayirli, E.; Mayes, A.; Eldridge, P.; Marson, A.G.; Wieshmann, U.C. A voxel-based asymmetry study of the relationship between hemispheric asymmetry and language dominance in Wada tested patients. *Hum. Brain Mapp.* **2018**, *39*, 3032–3045. [CrossRef]
16. Greve, D.N.; Van der Haegen, L.; Cai, Q.; Stufflebeam, S.; Sabuncu, M.R.; Fischl, B.; Brysbaert, M. A surface-based analysis of language lateralization and cortical asymmetry. *J. Cogn. Neurosci.* **2013**, *25*, 1477–1492. [CrossRef]
17. Josse, G.; Mazoyer, B.; Crivello, F.; Tzourio-Mazoyer, N. Left planum temporale: An anatomical marker of left hemispheric specialization for language comprehension. *Brain Res. Cogn. Brain Res.* **2003**, *18*, 1–14. [CrossRef]
18. Eckert, M.A.; Leonard, C.M.; Possing, E.T.; Binder, J.R. Uncoupled leftward asymmetries for planum morphology and functional language processing. *Brain Lang.* **2006**, *98*, 102–111. [CrossRef]

19. Dorsaint-Pierre, R.; Penhune, V.B.; Watkins, K.E.; Neelin, P.; Lerch, J.P.; Bouffard, M.; Zatorre, R.J. Asymmetries of the planum temporale and Heschl's gyrus: Relationship to language lateralization. *Brain* **2006**, *129*, 1164–1176. [CrossRef]
20. Jansen, A.; Liuzzi, G.; Deppe, M.; Kanowski, M.; Olschlager, C.; Albers, J.M.; Schlaug, G.; Knecht, S. Structural correlates of functional language dominance: A voxel-based morphometry study. *J. Neuroimaging* **2010**, *20*, 148–156. [CrossRef]
21. Luders, E.; Gaser, C.; Jancke, L.; Schlaug, G. A voxel-based approach to gray matter asymmetries. *Neuroimage* **2004**, *22*, 656–664. [CrossRef] [PubMed]
22. Kurth, F.; Luders, E.; Pigdon, L.; Conti-Ramsden, G.; Reilly, S.; Morgan, A.T. Altered gray matter volumes in language-associated regions in children with developmental language disorder and speech sound disorder. *Dev. Psychobiol.* **2018**, *60*, 814–824. [CrossRef] [PubMed]
23. Rosenberger, P.B.; Hier, D.B. Cerebral asymmetry and verbal intellectual deficits. *Ann. Neurol.* **1980**, *8*, 300–304. [CrossRef]
24. Qi, T.; Schaadt, G.; Friederici, A.D. Cortical thickness lateralization and its relation to language abilities in children. *Dev. Cogn. Neurosci.* **2019**, *39*, 100704. [CrossRef] [PubMed]
25. Koelkebeck, K.; Miyata, J.; Kubota, M.; Kohl, W.; Son, S.; Fukuyama, H.; Sawamoto, N.; Takahashi, H.; Murai, T. The contribution of cortical thickness and surface area to gray matter asymmetries in the healthy human brain. *Hum. Brain Mapp.* **2014**, *35*, 6011–6022. [CrossRef] [PubMed]
26. Zomorodian, A.; Carlsson, G. Computing persistent homology. *Discret. Comput. Geom.* **2005**, *33*, 249–274. [CrossRef]
27. Eckert, M.A.; Berninger, V.W.; Hoeft, F.; Vaden, K.I., Jr.; Dyslexia Data, C. A case of bilateral perisylvian syndrome with reading disability. *Cortex* **2016**, *76*, 121–124. [CrossRef]
28. Eckert, M.A.; Berninger, V.W.; Vaden, K.I., Jr.; Gebregziabher, M.; Tsu, L. Gray matter features of reading disability: A combined meta-analytic and direct analysis approach(1,2,3,4). *eNeuro* **2016**, 3. [CrossRef]
29. Eckert, M.A.; Vaden, K.I., Jr.; Maxwell, A.B.; Cute, S.L.; Gebregziabher, M.; Berninger, V.W.; Dyslexia Data, C. Common brain structure findings across children with varied reading disability profiles. *Sci. Rep.* **2017**, *7*, 6009. [CrossRef]
30. Eckert, M.A.; Vaden, K.I., Jr.; Roberts, D.R.; Castles, A.; Dyslexia Data, C. A pericallosal lipoma case with evidence of surface dyslexia. *Cortex* **2019**, *117*, 414–416. [CrossRef]
31. Manjon, J.V.; Coupe, P.; Marti-Bonmati, L.; Collins, D.L.; Robles, M. Adaptive non-local means denoising of MR images with spatially varying noise levels. *J. Magn. Reason. Imaging* **2010**, *31*, 192–203. [CrossRef]
32. Avants, B.B.; Tustison, N.J.; Song, G.; Cook, P.A.; Klein, A.; Gee, J.C. A reproducible evaluation of ANTs similarity metric performance in brain image registration. *Neuroimage* **2011**, *54*, 2033–2044. [CrossRef] [PubMed]
33. Kurth, F.; Gaser, C.; Luders, E. A 12-step user guide for analyzing voxel-wise gray matter asymmetries in statistical parametric mapping (SPM). *Nat. Protoc.* **2015**, *10*, 293–304. [CrossRef] [PubMed]
34. Joliot, M.; Jobard, G.; Naveau, M.; Delcroix, N.; Petit, L.; Zago, L.; Crivello, F.; Mellet, E.; Mazoyer, B.; Tzourio-Mazoyer, N. AICHA: An atlas of intrinsic connectivity of homotopic areas. *J. Neurosci. Methods* **2015**, *254*, 46–59. [CrossRef] [PubMed]
35. Brett, M.; Anton, J.L.; Valabregue, R.; Poline, J.B. Region of interest analysis using an SPM toolbox. In Proceedings of the 8th International Conference on Functional Mapping of the Human Brain, Sendai, Japan, 2–6 June 2002; p. 497.
36. Edelsbrunner, H.; Letscher, D.; Zomorodian, A. Topological persistence and simplification. In Proceedings of the 41st Annual Symposium on Foundations of Computer Science, Redondo Beach, CA, USA, 12–14 November 2000; pp. 454–463.
37. Heine, C.; Leitte, H.; Hlawitschka, M.; Iuricich, F.; De Floriani, L.; Scheuermann, G.; Hagen, H.; Garth, C. A survey of topology-based methods in visualization. *Comput Graph Forum* **2016**, *35*, 643–667. [CrossRef]
38. Tierny, J.; Favelier, G.; Levine, J.A.; Gueunet, C.; Michaux, M. The Topology ToolKit. *IEEE Trans. Vis. Comput. Graph.* **2018**, *24*, 832–842. [CrossRef]
39. Bubenik, P. Statistical topological data analysis using persistence landscapes. *J. Mach. Learn. Res.* **2015**, *16*, 77–102.

40. Fasy, B.T.; Kim, J.; Lecci, F.; Maria, C.; Millman, D.L.; Rouvreau, V. The included GUDHI is authored by Maria, C., Dionysus by Morozov, D., PHAT by Bauer, U., Kerber, M., Reininghaus, J. TDA: Statistical Tools for Topological Data Analysis. R Package Version 1.6.5. 2019. Available online: https://CRAN.R-project.org/package=TDA (accessed on 20 September 2020).
41. Rubin, D.B. Multiple imputation after 18+ years. *J. Am. Stat. Assoc.* **1996**, *91*, 473–489. [CrossRef]
42. Woodcock, R.W.; Mather, N.; McGrew, K.S.; Shrank, F.A. *Woodcock-Johnson III Tests of Cognitive Abilities*; Riverside Publishing: Itasca, IL, USA, 2001.
43. Woodcock, R. *Woodcock Reading Mastery Test: Revised*; American Guidance Service: Circle Pines, MN, USA, 1987.
44. Wagner, R.K.; Torgesen, J.K.; Rashotte, C.A. *Comprehensive Test of Phonological Processing*; Pro-Ed Inc.: Austin, TX, USA, 1999.
45. Wolf, M.; Denckla, M.B. *Rapid Automatized Naming and Rapid Alternating Stimulus Tests (RAN/RAS)*; Pro-Ed Inc.: Austin, TX, USA, 2005.
46. Wheeler, R.E.; Torchiano, M. ImPerm: Permutation Tests for Linear Models. R Package Version. 2010. Available online: https://CRAN.R-project.org/package=lmPerm (accessed on 20 September 2020).
47. Cohen, J.; Cohen, P. *Applied Multiple Regression/Correlation Analysis for the Behavioral Sciences*; Lawrence Erlbaum: New Jersey, NJ, USA, 1983.
48. Hothorn, T.; Kurt Hornik, K. exactRankTests: Exact Distributions for Rank and Permutation Tests. R Package Version 0.8-31. 2019. Available online: https://CRAN.R-project.org/package=exactRankTests (accessed on 20 September 2020).
49. Gaser, C.; Dahnke, R. CAT—A computational anatomy toolbox for the analysis of structural MRI data. In Proceedings of the Organization for Human Brain Mapping Annual Meeting, Geneva, Switzerland, 26–30 June 2016; pp. 336–348.
50. Smith, S.M.; Nichols, T.E. Threshold-free cluster enhancement: Addressing problems of smoothing, threshold dependence and localisation in cluster inference. *Neuroimage* **2009**, *44*, 83–98. [CrossRef]
51. Nunez, C.; Theofanopoulou, C.; Senior, C.; Cambra, M.R.; Usall, J.; Stephan-Otto, C.; Brebion, G. A large-scale study on the effects of sex on gray matter asymmetry. *Brain Struct. Funct.* **2018**, *223*, 183–193. [CrossRef]
52. Jahanshad, N.; Lee, A.D.; Barysheva, M.; McMahon, K.L.; de Zubicaray, G.I.; Martin, N.G.; Wright, M.J.; Toga, A.W.; Thompson, P.M. Genetic influences on brain asymmetry: A DTI study of 374 twins and siblings. *Neuroimage* **2010**, *52*, 455–469. [CrossRef] [PubMed]
53. Felton, A.; Vazquez, D.; Ramos-Nunez, A.I.; Greene, M.R.; McDowell, A.; Hernandez, A.E.; Chiarello, C. Bilingualism influences structural indices of interhemispheric organization. *J. Neurolinguistics* **2017**, *42*, 1–11. [CrossRef] [PubMed]
54. Corballis, M.C. Humanity and the left hemisphere: The story of half a brain. *Laterality* **2020**, 1–15. [CrossRef] [PubMed]
55. Neubauer, S.; Gunz, P.; Scott, N.A.; Hublin, J.J.; Mitteroecker, P. Evolution of brain lateralization: A shared hominid pattern of endocranial asymmetry is much more variable in humans than in great apes. *Sci. Adv.* **2020**, *6*, eaax9935. [CrossRef]

Publisher's Note: MDPI stays neutral with regard to jurisdictional claims in published maps and institutional affiliations.

Article

Rightward Shift of Two-Channel NIRS-Defined Prefrontal Cortex Activity during Mental Arithmetic Tasks with Increasing Levels of State Anxiety

Miwa Horiuchi-Hirose [1,*] **and Kazuhiko Sawada** [2]

1 Department of Nursing, Ibaraki Christian University, Hitachi, Ibaraki 319-1295, Japan
2 Department of Nutrition, Faculty of Medical and Health Sciences, Tsukuba International University, Tsuchiura, Ibaraki 300-0051, Japan; k-sawada@tius.ac.jp
* Correspondence: hirose@icc.ac.jp; Tel.: +81-52-3215

Received: 12 February 2020; Accepted: 23 March 2020; Published: 3 April 2020

Abstract: This study was aimed at clarifying the effect of different levels of state anxiety caused by mental arithmetic tasks on the anxiety- and/or task performance-related activation of the frontopolar prefrontal cortex (PFC). Twenty-six healthy male subjects performed two sets of mental arithmetic tasks, which consisted of two difficulty levels. Anxiety levels were evaluated subjectively by the State–Trait Anxiety Inventory-Form JYZ (STAI). Near-infrared spectroscopy (NIRS) measurements revealed greater levels of oxyhemoglobin in the frontopolar PFC during experimental tasks. When the subjects were divided into three anxiety groups based on STAI scores, arithmetic task performance was reduced in the moderate and high state anxiety groups compared the low state anxiety group during the experimental task, but not in the control task. Increased frontopolar PFC activity during the experimental task was observed on either side in the moderate anxiety group. The laterality of frontopolar PFC activity in moderate and high state anxiety groups shifted from left to right dominance, independent of task difficulty. Our findings suggested that reduced task performance increased the difficulty of the arithmetic tasks and was involved in the state anxiety-associated rightward lateralization of the frontopolar PFC.

Keywords: oxyhemoglobin level; state anxiety; task performance; heart rate; human

1. Introduction

The prefrontal cortex (PFC) undergoes remarkably slow maturation and is probably the last brain region to achieve cortical myelination during adolescence and adulthood [1,2]. Recent human imaging studies have demonstrated that various cognitive tasks activate the PFC subdivisions, such as the frontopolar or rostral PFC (Brodmann's area (BA) 10), dorsolateral PFC (BA 9/46) and ventrolateral PFC (BA 12/45 and lower 46) [3–6]. The BA 10, which presents the most extensive cytoarchitectural region of the human cerebral cortex, is located on the most anterior part of the frontal lobe, comprising the rostral prefrontal (or frontopolar) cortex [7], and is involved in task complexity [8]. The BA10 plays an important role in the handling of unpredictable tasks [9] and is the seat of working memory [10]. On the other hand, the asymmetric aspects of the PFC function have been documented. The PFC is known to localize to the left hemisphere for positive experiences and to the right hemisphere for negative experiences [11]. The left medial orbitofrontal cortex was correlated positively with rewards [12]. There are associations that were reported regarding the frontal activity on the left side with approach-oriented emotional states (anger or joy) and on the right side with withdrawal-oriented emotional states (fear or sadness) [13,14]. Furthermore, the relatively left-dominant frontal activity reflected greater motivation approach and positive effects, and the relatively right-dominant frontal activity reflected greater motivation withdrawal and negative effects [15].

Recent neuroimaging studies revealed the neural basis of PFC roles in the mood-cognition interaction using a verbal working memory task [16–18] and the n-back task [19]. Similar associations have been observed for some psychiatric disorders, such as generalized anxiety disorder, psychosis, and mood disorders [20–22].

Anxiety is an aversive emotional and motivational state that occurs in threatening circumstances [23]. The State–Trait Anxiety Inventory-Form JYZ (STAI) [24] is a self-report test evaluating the presence and severity of current symptoms of anxiety. The state anxiety scale from the STAI evaluates sensitive situational changes reflected by fear, tension, nervousness, and troubled thinking. State anxiety has been conceptualized as a transient emotional state during which an individual is unable to respond to events [23]. Anxiety during continuous arithmetic tasks was associated with near-infrared spectroscopy (NIRS)-defined activity in the dorsolateral (BA9) and frontopolar (BA10) PFC [25]. Hemodynamic responses occurring during state and trait anxiety were different [25]. However, the arithmetic task performances altered hemodynamic responses in the ventrolateral PFC (BA44, BA45, and BA47), without having direct correlations with state anxiety-related PFC activation during continuous arithmetic tasks [25]. We focused on the frontopolar PFC (BA10), which is known to be activated by state anxiety [3,5] in relation to anxiety under maintained psychological stress [25]. The present study aimed to clarify the effect of differential levels of state anxiety during mental arithmetic tasks on the anxiety- and/or task performance-related activation of the PFC. Two-channel NIRS was used for evaluating the activity of the PFC subdivision corresponding to the frontopolar cortex covered by Fp1 and Fp2 settings, in accordance with 10–20 international systems for electroencephalography. We obtained intriguing findings about state anxiety-related changes in the laterality of frontopolar PFC activity during mental arithmetic tasks.

2. Materials and Methods

2.1. Participants

The participants in this study were university students, who were recruited using both the Ibaraki Christian University website and flyers that were distributed throughout the university campus. Participants received a reward for participation. In total, 26 Japanese male university students (mean age: 21.0 ± 1.1 years in mean BMI: 21.3 ± 2.6) participated in this study. All subjects were screened for histories of neurological, psychiatric, and cardiovascular disease, using a self-report questionnaire. Dominant hands were evaluated by the Edinburgh Handedness Inventory [26]. Twenty-two subjects were right-handed and four were left-handed. Written informed consent was obtained before commencing the study. This study was approved by the ethics committee at Ibaraki Christian University (code number 2016-020).

2.2. Measures

2.2.1. Task

All subjects rested in the laboratory for 10 minutes before the tasks. The mental arithmetic tasks, as psychological stressors, were used in reference to the studies by Wang et al. [27] and Dedovic et al. [28]. The subtraction of two-digit numbers from four-digit numbers were defined as experimental arithmetic tasks, and the subtraction of single digit numbers from two-digit numbers were defined as control arithmetic tasks. After receiving instructions concerning the task procedures, subjects carried out two sample trials of the tasks. Then, subjects performed two sets of 12 consecutive trials of the arithmetic tasks, shown in black letters on a white background on PowerPoint slides. The arithmetic task sets consisted of experimental and control tasks. The first task was randomly selected. Subjects were given 10 seconds for each trial, followed by the display of the correct answers of each trial. Subjects could use the clicker (ResponseCard LT, Keepad Japan, Osaka, Japan) on the monitor to choose from four options on each trial. Task performances, i.e, the number of correct responses, the number of error responses, and the reaction time, were recorded and analyzed with Turning Point software (Keepad Japan, Osaka, Japan). Just before starting the first task, the fixation mark (+ mark) was shown on the monitor for five seconds to measure the baseline ECG(electrocardiogram) for calculating the cardiological index. The State–Trait Anxiety Inventory-Form JYZ (STAI) is a self-reporting form for evaluating the presence and severity of current anxiety symptoms. Subjects filled out STAI questionnaires (state anxiety and trait anxiety) before the first task set and the results were defined as the rest condition. STAI questionnaires given in the intervals between the first and second task sets were classified into two conditions, depending on the difficulty of the first task set. The control condition was defined when the subjects performed the control task as the first task set. The experimental condition was defined when the subjects performed the experimental task as the first task set. The control and experimental conditions were defined when the subjects performed the control and experimental tasks as the first task sets, respectively. In order to evaluate the effects of state anxiety levels on task performances, cardiological responses and the PFC activity in the experimental and control conditions, subjects with low anxiety in the control condition were further divided into three classes of state anxiety levels in the experimental task based on STAI scores: low state anxiety level (STAI scores < 45), moderate state anxiety level (45 < STAI scores < 55) and high state anxiety level (STAI scores > 55), depending on the arithmetic task types [24].

2.2.2. ECG Measurements

Heart rate (HR) was measured as a reference for autonomic nervous activity. The R-R intervals were measured by a heart-rate sensing system (Poral V800, Poral Japan, Tokyo, Japan). The ECG measurements were performed just before starting the first and second tasks (when the fixation mark (+ mark) was shown to subjects for five seconds as baseline measurements), and during the first and second tasks. The results were expressed as the percentage of changes in HR during the first and second tasks from each baseline value.

2.2.3. NIRS Measurements

NIRS measurements were carried out using the two-channel wireless system (Pocket NIRS HM, DynaSence, Hamamatsu, Japan). This tool measured changes in levels of oxygenated hemoglobin (oxyhemoglobin–hemoglobin) and deoxygenated hemoglobin (deoxyhemoglobin) at wavelengths of 735 and 850 nm. Concentrations of oxyhemoglobin–hemoglobin and deoxyhemoglobin levels were based on the modified Beer–Lambert law. Data were recorded at sampling rates of 30 Hz along an Fp1 to Fp2 line 10–20 electroencephalogram electrode placement system, covering both the left and right sides of the frontopolar PFC (BA 10) [29]. For evaluation of the left/right side bias of the PFC activity, the Laterality Index (LI) was calculated using the formula (ΔRight oxyhemoglobin–hemoglobin—Δ Left oxyhemoglobin–hemoglobin)/(ΔRight oxy -hemoglobin +ΔLeft oxyhemoglobin–hemoglobin) [30]. LI values indicated the direction of asymmetry (negative values = left dominance; positive values = right dominance). Because LI values show negative numbers when the direction of asymmetry is left-dominant, mean LI values will be lower if many individuals show left-dominant asymmetry and will sometimes be negative.

2.3. Statistical Analysis

Significant differences in task performances and cardiological index between the experimental and control task conditions were evaluated using paired *t*-tests. We evaluated correlations between state anxiety and trait anxiety scores using a correlated sample *t*-test. We used two-way repeated measures ANOVA to detect changes in oxyhemoglobin–hemoglobin levels and deoxy–hemoglobin levels in the PFC using NIRS channels (left and right PFC) and task conditions (experimental and control tasks) as factors, followed by Scheffe's test for post-hoc testing. Subsequently, task performance scores, cardiological index, and oxyhemoglobin levels in the PFC were compared among three levels of state anxiety groups divided on the basis of STAI scores—that is, low, moderate, and high state anxiety groups. Two-way repeated measures ANOVA was carried out to evaluate changes in task performance scores, cardiological index, and LI scores using state anxiety levels (low, moderate, and high) and task conditions (experimental and control tasks) as factors. A paired *t*-test was carried out as post-hoc testing to access significant difference in each measurement/index between experimental and control conditions, and Scheffe's test was carried out to access significant differences among three anxiety groups. Furthermore, we used repeated measures three-way ANOVA for detecting task condition- and anxiety level-related changes in oxyhemoglobin–hemoglobin level laterality in the frontopolar PFC. Paired t-tests were carried out to access significant difference in oxyhemoglobin–hemoglobin levels between left/right sides of the PFC or between experimental and control conditions. Scheffe's test was carried out to assess significant differences in the oxyhemoglobin levels in the left and right PFC among the three anxiety groups.

3. Results

3.1. STAI Scores, Task Performances and Cardiological Index

The trait anxiety scores were 46.2 ± 8.4, showing a significantly positive correlation with state anxiety at the resting condition. The state anxiety scores after the experimental condition (45.6 ± 13.2) was significantly higher than those at rest (39.3 ± 9.0, $p < 0.05$), but not after the control condition (35.8 ± 12.1, $p = 0.113$). Task performance scores such as the numbers of correct responses, error responses, and the reaction time were compared between the experimental and control tasks. The number of correct responses during the experimental task (5.3 ± 2.3) was significantly smaller than that during the control task (11.7 ± 0.7, $p < 0.001$) (Table 1). Reciprocally, the number of error responses was significantly higher during the experimental task (3.6 ± 2.6) than during the control task (0.1 ± 0.3, $p < 0.001$) (Table 1). A significant longer reaction time was obtained in the experimental task (8.1 ± 0.9) than in the control task (2.2 ± 0.5, $p < 0.001$) (Table 1). Thus, mental arithmetic task performances reduced with increasing levels of task difficulty (experimental task).

Table 1. Mental arithmetic task performance scores and cardiological index of all subjects.

	n	CONT Task			EX Task			*P* Value
Task performance scores								
Number of correct responses	26	11.7	±	0.7	5.3	±	2.3	0.000 **
Number of error responses	26	0.1	±	0.3	3.6	±	2.6	0.000 **
Reaction time	26	2.2	±	0.5	8.1	±	0.9	0.000 **
Cardiological index								
HR (% of change during task sets)	24	100.1	±	7.3	103.5	±	8.9	0.167

Control (CONT); experimental (EX); heart rate (HR). ** $p < 0.001$ at determined using paired *t*-test.

The heart rates (HR) at baseline levels were not significantly different between experimental and control tasks. The percentage changes of each cardiological index from the baseline levels were not significantly different between subjects who performed the experimental and control tasks (Table 1). Autonomic nervous activity was not altered by the difficulty of mental arithmetic tasks.

3.2. Oxyhemoflobin Levels

The frontopolar PFC activity during mental arithmetic tasks was evaluated by measuring oxy-and deoxyhemoglobin levels using NIRS. Figure 1 shows bar graphs of oxy- and deoxyhemoglobin levels during experimental and control tasks. Repeated measures two-way ANOVA was performed using the left/right sides of hemispheres and task conditions as factors. A significant effect on oxyhemoglobin levels was detected in task conditions [$F_{(1,25)} = 19.832$, $p < 0.0001$], but not in the left/right sides, and an interaction between task conditions and the left/right sides of hemispheres. Post-hoc testing revealed a significantly greater oxyhemoglobin levels in the ipsilateral PFCs during the experimental task than during the control tasks ($p < 0.001$). Thus, the frontopolar PFC was activated bilaterally by increasing levels of mental arithmetic task difficulty.

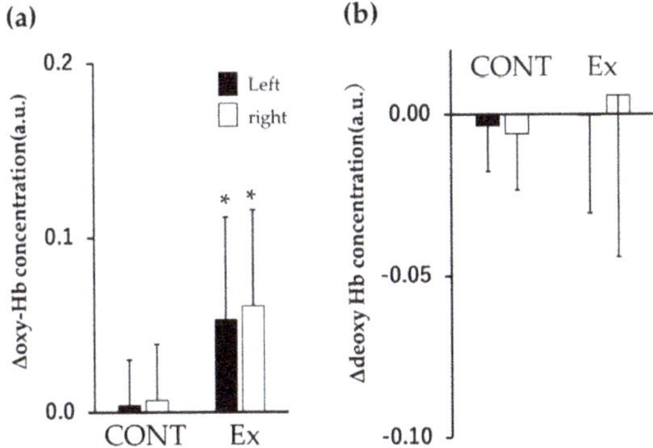

Figure 1. Bar graphs of near-infrared spectroscopy (NIRS)-defined oxyhemoglobin and deoxyhemoglobin levels. (**a**) Oxyhemoglobin concentration during experimental and control tasks. (**b**) deoxyhemoglobin concentration during experimental and control tasks. n = 26, mean ± SD, * $p < 0.001$ (experimental task vs. control task; Scheffe's test). Control (CONT); experimental (EX).

3.3. *State Anxiety Levels Following Mental Arithmetic Tasks*

Subjects in the experimental and control conditions were further divided into three classes of state anxiety levels based on STAI scores: low, moderate, and high levels. A great majority of the subjects in the control condition (92.3%; 24 out of 26 subjects) exhibited low-level state anxiety. The remaining two subjects showed high levels of anxiety. When the subjects with low anxiety in the control condition performed the experimental task, state anxiety levels were low in 41.7% (10 out of 24 subjects), moderate in 45.8% (11 out of 24 subjects), and high in 12.5% (three out of 24 subjects). They were defined as the EX/low, EX/moderate and EX/high anxiety groups, respectively.

3.3.1. Task Performance Scores

Task performance scores were compared among three groups of different state anxiety levels, and between task conditions in each group (Table 2). Repeated measures two-way ANOVA revealed significant effects of task conditions [$F_{(1,21)} = 381.966$, $p < 0.001$] and the interaction between task conditions and anxiety levels [$F_{(2,21)} = 11.210$, $p < 0.05$] on the number of correct responses. Post-hoc testing indicated that a significantly greater number of correct responses was given in the control task than in the experimental task in all three state anxiety groups. The number of correct responses reduced significantly in the EX/moderate and EX/high anxiety groups when compared to the EX/low anxiety group (Table 2). Significant effects were obtained by repeated measures two-way ANOVA for task conditions on the number of incorrect responses [$F_{(1,21)} = 121.002$, $p < 0.001$] and on the reaction time [$F_{(1,21)} = 289.296$, $p < 0.05$]. Although a significantly frequent number of error responses and significantly longer reaction times were detected in the experimental task compared to the control task in all three state anxiety groups by paired *t*-test, there were no differences in these two task scores among the three anxiety groups by Scheffe's test (Table 2). Thus, task performances were reduced by increasing levels of the difficulty of mental arithmetic tasks. The anxiety levels were altered in accordance with a decreased number of correct responses, but did not affect other task performances.

Table 2. Results of task performances, cardiological indexes, and NIRS-defined oxyhemoglobin levels in EX/low, EX/moderate and EX/high state anxiety groups.

	CONT task			EX task			vs. EX/Low (Scheffe's Test)
EX/Low anxiety group (n = 10)							
Task performance scores							
Number of correct responses	11.4	±	1.1	6.6	±	2.7 **	–
Number of error response	0.2	±	0.4	2.1	±	1.6 *	–
Reaction time	2.1	±	0.4	7.5	±	1.2 **	–
Cardiological index							
HR (% change during task sets)	101.7	±	4.8	102.2	±	5.2	–
Oxyhemoglobin levels							
Left hemisphere	0.007	±	0.013	0.034	±	0.070	–
Right hemisphere	0.018	±	0.028	0.048	±	0.045	–
Laterality Index (LI)	−0.989	±	3.719	−0.311	±	0.583	–
EX/Moderate anxiety group (n = 11)							
Task performance scores							
Number of correct responses	12.0	±	0	4.5	±	1.6 **	$p < 0.025$
Number of error response	0	±	0	4.4	±	3.1 **	NS
Reaction time	2.3	±	0.6	8.5	±	0.4 **	NS
Cardiological index							
HR (% change during task sets)	97.9	±	6.7	105.8	±	10.8	NS
Oxyhemoglobin levels							
Left hemisphere	−0.002	±	0.034	0.057	±	0.049 **	NS
Right hemisphere	−0.002	±	0.036	0.076	±	0.072 **	NS
Laterality Index (LI)	0.135	±	0.772	0.177	±	0.437	NS
EX/High anxiety group (n = 3)							
Task performance scores							
Number of correct responses	12.0	±	0	4.3	±	1.5 *	$p < 0.05$
Number of error response	0	±	0	5.0	±	1.7 *	NS
Reaction time	2.4	±	0.6	8.1	±	0.7 **	NS
Cardiological index							
HR (% change during task sets)	98.7	±	9.5	105.1	±	13.1	NS
Oxyhemoglobin levels							
Left hemisphere	0.028	±	0.015	0.054	±	0.028	NS
Right hemisphere	0.018	±	0.002	0.057	±	0.016 *	NS
Laterality Index (LI)	−0.129	±	0.376	0.049	±	0.327	NS

Heart rate (HR); high frequency (HF); low frequency (LF); control (CONT); experimental (EX).* $p < 0.05$, ** $p < 0.025$ (EX vs. CONT task) paired *t-test*.

3.3.2. Cardiological Index

The HR was not altered by the state anxiety level or the difficulty of the mental arithmetic tasks. No significant effect on HR was observed for any group for state anxiety levels or task conditions, as assessed by repeated measures two-way ANOVA (Table 2).

3.3.3. Oxyhemoglobin Levels

The oxyhemoglobin levels of the left and right cerebral hemispheres were calculated as the frontopolar PFC activity, and were compared between the task conditions in all three groups of state anxiety levels (Table 2). Repeated measures three-way ANOVA revealed significant effects on task conditions [$F_{(1,84)} = 0.032$, $p < 0.001$]. Post-hoc testing indicated that there were significantly greater oxyhemoglobin levels associated with experimental tasks compared to the control task, which were obtained in both sides of cerebral hemispheres of EX/moderate group, and only in the right hemisphere in EX/high anxiety group (Table 2). Significant left/right side differences in oxyhemoglobin levels were not detected at a population level in either condition of all state anxiety groups (Table 2).

The left/right side differences in oxyhemoglobin levels at an individual level were evaluated by the LI. Although there were no differences in the mean LI between the task conditions and among state anxiety groups by repeated measures two-way ANOVA, the LI revealed that the laterality of the frontopolar PFC activity changed depending on the state anxiety level at an individual level. The strongest leftward laterality of the frontopolar PFC activity was seen in the EX/low anxiety group (Table 2). With increasing state anxiety levels, the laterality of frontopolar PFC activity shifted from left to right dominance independent of task difficulty.

4. Discussion

Arithmetic task performances activated the superior and middle frontal gyri (identical to the frontal polar cortex; BA10) in difficult tasks and the inferior frontal gyrus in easy tasks [31]. On the other hand, Takizawa et al. [25] reported that the frontopolar PFC was associated with state anxiety during arithmetic tasks. Our results suggest that both the left and right sides of the frontopolar PFC was activated by increasing the difficulty of mental arithmetic tasks and its activity shifted with increasing anxiety levels. State anxiety was found to have no direct association with arithmetic task performances [25]. It was unclear whether state anxiety reduced arithmetic task performance by itself in the present study. However, the task performance reduction with increasing arithmetic task difficulty enhanced the state anxiety-related rightward shift of the frontopolar PFC activity caused by state anxiety. These findings suggest that state anxiety acts as a modulator facilitating the reduced performance associated with mental arithmetic tasks.

The PFC activity in the resting state was right-dominant in subjects with higher trait anxiety levels, but left-dominant in subjects with lower trait anxiety levels [32]. Increased activity of the right hemisphere, including the PFC, is also found in patients with major depressive disorder (MDD) at the resting state [33]. Although a significant positive correlation was shown between STAI-defined trait and state anxiety scores at resting conditions, the lateralized PFC activity during the arithmetic task performance in the present study was associated with state anxiety levels, but not with trait anxiety levels. On the other hand, MDD patients had reduced left PFC activity during verbal fluency tasks (VFT) [34] and emotionally challenging tasks [15]. The left PFC activity in MDD patients was reduced during cognitive or emotional tasks by impairing the downregulation of amygdala responses to negative emotional information [35]. Patients with anxiety disorders exhibit reduced left hemisphere activity for syllables compared to those without anxiety disorders [36]. In our results, the state anxiety-related lateralization of PFC activity may be involved in the rightward shift of PFC activity rather than a reduction in left PFC activity.

The brain regions related to cognition and emotion are known to lateralize morphologically and functionally [37–39]. Children with math anxiety were shown to have elevated connectivity between the amygdala and ventromedial PFC regions [40]. The cerebral blood flow in the ventral PFC was induced during arithmetic tasks under high stress levels, and sustained at the right side, but not the left side, after task completion [27]. Functional NIRS revealed that state anxiety levels were negatively correlated with right-predominant activation of the ventrolateral and orbital PFC during auditory working memory tasks [41]. Thus, the ventral and ventrolateral regions of the PFC were activated at the right side by state anxiety during various tasks. The right cerebral hemisphere displayed negative feedback circuits for the inhibitory processes of cognitive, affective, and physiological regulations [42]. Since ventrolateral PFC activation was associated with arithmetic task performance, rather than state anxiety [25], there may be negative feedback circuits connecting the frontopolar PFC and ventral/ventrolateral PFC for reducing arithmetic task performance. State anxiety may evoke such negative feedback circuits by shifting frontopolar PFC activity in a rightward manner.

The orbital and medial PFC both inhibited the amygdala [43]. The cerebral blood flow baseline on the right side of these two PFC subdivisions was variable in correlation with changing salivary cortisol levels and HR during mental arithmetic tasks [27]. These findings suggest that the orbital and medial PFC may be involved in HR control. Mathematics anxiety also caused strongly influenced by HR responses compared to the anxiety evoked by the Speech and Stroop test [44]. In the present study, HR was not altered by either state anxiety levels or arithmetic task difficulty, suggesting that sympathetic HR activity associated with psychological math stress is tightly linked with orbital and medial PFC activity, rather than frontopolar PFC activity.

The present study evaluated hemodynamic responses in the frontopolar PFC, but not in other brain regions, such as the ventrolateral and orbital PFC, during mental arithmetic tasks using two-channel NIRS. Although we focused on limited subdivisions of the PFC, it was revealed that state anxiety shifted toward right dominance in the frontopolar PFC activity, which reduced mental arithmetic task performances. Mathematics anxiety levels were evaluated using several variables, such as ability, school grade levels, and undergraduate fields of study [45]. Anxiety may not impair the performance effectiveness (quality of performance) when it leads to the use of compensatory strategies [15]. In the future, it will be necessary to examine compensatory strategies for modulating mental arithmetic task performances against state anxiety-related lateralized activation of the frontopolar PFC and other PFC subdivisions.

The major limitation of this study was that the use of two-channel NIRS resulted in a low spatial resolution, which limited the studied region of interest to PFC subdivisions. Other imaging techniques with high spatial resolution, such as positron emission tomography and functional magnetic resonance imaging (fMRI), may further refine the PFC subdivisions that are associated with math stress-induced state anxiety. Further studies will be necessary to determine which brain regions are involved. Another limitation of this study was that the presence and severity of anxiety symptoms were evaluated using a self-report scale, the STAI. A larger sample size may improve the accuracy of subjective evaluations of state anxiety levels. Furthermore, the EX/high level anxiety group only contained three subjects. A larger sample size may also improve the reliability of statistical analysis.

Author Contributions: M.H.-H. designed the study and collected data; M.H.-H. and K.S. analyzed and wrote the first draft of the manuscript. All authors approved the final version of the manuscript.

Funding: This research was funded by JSPS KAKEN, Grant number 16K20730.

Acknowledgments: The authors would like to thank Enago (www.enago.jp) for the English language review.

Conflicts of Interest: The authors declare no conflict of interest.

References

1. Sowell, E.R.; Thompson, P.M.; Holmes, C.J.; Jernigan, T.L.; Toga, A.W. In vivo evidence for post-adolescent brain maturation in frontal and striatal regions. *Nat. Neurosci.* **1999**, *2*, 859–861. [CrossRef] [PubMed]
2. Sowell, E.R.; Thompson, P.M.; Leonard, C.M.; Welcome, S.E.; Kan, E.; Toga, A.W. Longitudinal mapping of cortical thickness and brain growth in normal children. *J. Neurosci.* **2004**, *24*, 8223–8231. [CrossRef] [PubMed]
3. Bishop, S.; Duncan, J.; Brett, M.; Lawrence, A.D. Prefrontal cortical function and anxiety: Controlling attention to threat-related stimuli. *Nat. Neurosci.* **2004**, *7*, 184–188. [CrossRef] [PubMed]
4. MacDonald, A.W., 3rd; Cohen, J.D.; Stenger, V.A.; Carter, C.S. Dissociating the role of the dorsolateral prefrontal and anterior cingulate cortex in cognitive control. *Science* **2000**, *288*, 1835–1838. [CrossRef] [PubMed]
5. Duncan, J.; Owen, A.M. Common regions of the human frontal lobe recruited by diverse cognitive demands. *Trends Neurosci.* **2000**, *23*, 475–483. [CrossRef]
6. Botvinick, M.; Nystrom, L.E.; Fissell, K.; Carter, C.S.; Cohen, J.D. Conflict monitoring versus selection-for-action in anterior cingulate cortex. *Nature* **1999**, *402*, 179–181. [CrossRef]
7. Ramnani, N.; Owen, A.M. Anterior prefrontal cortex: Insights into function from anatomy and neuroimaging. *Nat. Rev. Neurosci.* **2004**, *5*, 184–194. [CrossRef]
8. Chahine, G.; Diekhof, E.K.; Tinnermann, A.; Gruber, O. On the role of the anterior prefrontal cortex in cognitive 'branching': An fMRI study. *Neuropsychologia* **2015**, *77*, 421–429. [CrossRef]
9. Koechlin, E.; Corrado, G.; Pietrini, P.; Grafman, J. Dissociating the role of the medial and lateral anterior prefrontal cortex in human planning. *Proc. Natl. Acad. Sci. USA* **2000**, *97*, 7651–7656. [CrossRef]
10. Braver, T.S.; Bongiolatti, S.R. The role of frontopolar cortex in subgoal processing during working memory. *Neuroimage* **2002**, *15*, 523–536. [CrossRef]
11. Jones, N.A.; Fox, N.A. Electroencephalogram asymmetry during emotionally evocative films and its relation to positive and negative affectivity. *Brain Cogn.* **1992**, *20*, 280–299. [CrossRef]
12. O'Doherty, J.; Kringelbach, M.L.; Rolls, E.T.; Hornak, J.; Andrews, C. Abstract reward and punishment representations in the human orbitofrontal cortex. *Nat. Neurosci.* **2001**, *4*, 95–102. [CrossRef] [PubMed]
13. Coan, J.A.; Allen, J.J. Frontal EEG asymmetry as a moderator and mediator of emotion. *Biol. Psychol.* **2004**, *67*, 7–49. [CrossRef] [PubMed]

14. Harmon-Jones, E.; Sigelman, J. State anger and prefrontal brain activity: Evidence that insult-related relative left-prefrontal activation is associated with experienced anger and aggression. *J. Personal. Soc. Psychol.* **2001**, *80*, 797–803. [CrossRef]
15. Harmon-Jones, E.; Gable, P.A.; Peterson, C.K. The role of asymmetric frontal cortical activity in emotion-related phenomena: A review and update. *Biol. Psychol.* **2010**, *84*, 451–462. [CrossRef] [PubMed]
16. Aoki, R.; Sato, H.; Katura, T.; Utsugi, K.; Koizumi, H.; Matsuda, R.; Maki, A. Relationship of negative mood with prefrontal cortex activity during working memory tasks: An optical topography study. *Neurosci. Res.* **2011**, *70*, 189–196. [CrossRef]
17. Aoki, R.; Sato, H.; Katura, T.; Matsuda, R.; Koizumi, H. Correlation between prefrontal cortex activity during working memory tasks and natural mood independent of personality effects: An optical topography study. *Psychiatry Res.* **2013**, *212*, 79–87. [CrossRef]
18. Sato, H.; Dresler, T.; Haeussinger, F.B.; Fallgatter, A.J.; Ehlis, A.C. Replication of the correlation between natural mood states and working memory-related prefrontal activity measured by near-infrared spectroscopy in a German sample. *Front. Hum. Neurosci.* **2014**, *8*, 37. [CrossRef]
19. Ozawa, S.; Matsuda, G.; Hiraki, K. Negative emotion modulates prefrontal cortex activity during a working memory task: A NIRS study. *Front. Hum. Neurosci.* **2014**, *8*, 46. [CrossRef]
20. Han, K.M.; De Berardis, D.; Fornaro, M.; Kim, Y.K. Differentiating between bipolar and unipolar depression in functional and structural MRI studies. *Prog. Neuropsychopharmacol. Biol. Psychiatry* **2019**, *91*, 20–27. [CrossRef]
21. Tempesta, D.; Mazza, M.; Serroni, N.; Moschetta, F.S.; Di Giannantonio, M.; Ferrara, M.; De Berardis, D. Neuropsychological functioning in young subjects with generalized anxiety disorder with and without pharmacotherapy. *Prog. Neuropsychopharmacol. Biol. Psychiatry* **2013**, *45*, 236–241. [CrossRef] [PubMed]
22. Ebisch, S.J.; Mantini, D.; Northoff, G.; Salone, A.; De Berardis, D.; Ferri, F.; Ferro, F.M.; Di Giannantonio, M.; Romani, G.L.; Gallese, V. Altered brain long-range functional interactions underlying the link between aberrant self-experience and self-other relationship in first-episode schizophrenia. *Schizophr. Bull.* **2014**, *40*, 1072–1082. [CrossRef] [PubMed]
23. Eysenck, M.W.; Derakshan, N.; Santos, R.; Calvo, M.G. Anxiety and cognitive performance: Attentional control theory. *Emotion* **2007**, *7*, 336–353. [CrossRef] [PubMed]
24. Hidano, T.; Fukuhara, M.; Iwawaki, M.; Soga, S.; Spielberger, C. *State Trait Anxiety Inventory (Form JYZ). Test Manual. (Japanese Adaptation of STAI)*; Jitsumu Kyouiku Shuppan: Tokyo, Japan, 2000.
25. Takizawa, R.; Nishimura, Y.; Yamasue, H.; Kasai, K. Anxiety and performance: The disparate roles of prefrontal subregions under maintained psychological stress. *Cereb. Cortex* **2014**, *24*, 1858–1866. [CrossRef] [PubMed]
26. Salmaso, D.; Longoni, A.M. Problems in the assessment of hand preference. *Cortex* **1985**, *21*, 533–549. [CrossRef]
27. Wang, J.; Rao, H.; Wetmore, G.S.; Furlan, P.M.; Korczykowski, M.; Dinges, D.F.; Detre, J.A. Perfusion functional MRI reveals cerebral blood flow pattern under psychological stress. *Proc. Natl. Acad. Sci. USA* **2005**, *102*, 17804–17809. [CrossRef]
28. Dedovic, K.; Renwick, R.; Mahani, N.K.; Engert, V.; Lupien, S.J.; Pruessner, J.C. The montreal imaging stress task: Using functional imaging to investigate the effects of perceiving and processing psychosocial stress in the human brain. *J. Psychiatry Neurosci.* **2005**, *30*, 319–325.
29. Okamoto, M.; Dan, H.; Sakamoto, K.; Takeo, K.; Shimizu, K.; Kohno, S.; Oda, I.; Isobe, S.; Suzuki, T.; Kohyama, K.; et al. Three-dimensional probabilistic anatomical cranio-cerebral correlation via the international 10–20 system oriented for transcranial functional brain mapping. *Neuroimage* **2004**, *21*, 99–111. [CrossRef]
30. Davidson, R.J.; Fox, N.A. Asymmetrical brain activity discriminates between positive and negative affective stimuli in human infants. *Science* **1982**, *218*, 1235–1237. [CrossRef]
31. Verner, M.; Herrmann, M.J.; Troche, S.J.; Roebers, C.M.; Rammsayer, T.H. Cortical oxygen consumption in mental arithmetic as a function of task difficulty: A near-infrared spectroscopy approach. *Front. Hum. Neurosci.* **2013**, *7*, 217. [CrossRef]
32. Ishikawa, W.; Sato, M.; Fukuda, Y.; Matsumoto, T.; Takemura, N.; Sakatani, K. Correlation between asymmetry of spontaneous oscillation of hemodynamic changes in the prefrontal cortex and anxiety levels: A near-infrared spectroscopy study. *J. Biomed. Opt.* **2014**, *19*, 027005. [CrossRef] [PubMed]

33. Li, M.; Xu, H.; Lu, S. Neural basis of depression related to a dominant right hemisphere: A resting state fMRI study. *Behav. Neurol.* **2018**, *2018*, 5024520. [CrossRef] [PubMed]
34. Baik, S.Y.; Kim, J.Y.; Choi, J.; Baek, J.Y.; Park, Y.; Kim, Y.; Jung, M.; Lee, S.H. Prefrontal asymmetry during cognitive tasks and its relationship with suicide ideation in major depressive disorder: An fNIRS study. *Diagnostics* **2019**, *9*, 193. [CrossRef] [PubMed]
35. Bruder, G.E.; Stewart, J.W.; McGrath, P.J. Right brain, left brain in depressive disorders: Clinical and theoretical implications of behavioral, electrophysiological and neuroimaging findings. *Neurosci. Biobehav. Rev.* **2017**, *78*, 178–191. [CrossRef] [PubMed]
36. Bruder, G.E.; Alvarenga, J.; Abraham, K.; Skipper, J.; Warner, V.; Voyer, D.; Peterson, B.S.; Weissman, M.M. Brain laterality, depression and anxiety disorders: New findings for emotional and verbal dichotic listening in individuals at risk for depression. *Laterality* **2016**, *21*, 525–548. [CrossRef] [PubMed]
37. Alves, N.T.; Fukusima, S.S.; Aznar-Casanova, J.A. Models of brain asymmetry in emotional processing. *Psychol. Neurosci.* **2008**, *1*, 63–66. [CrossRef]
38. Hu, D.; Shen, H.; Zhou, Z. Functional asymmetry in the cerebellum: A brief review. *Cerebellum* **2008**, *7*, 304–313. [CrossRef]
39. Ashwell, K.; Mai, J. Fetal development of the central nervous system. In *The Human Nervous System*, 3rd ed.; Elsevier Academic Press: San Diego, CA, USA, 2012.
40. Moustafa, A.A.; Tindle, R.; Ansari, Z.; Doyle, M.J.; Hewedi, D.H.; Eissa, A. Mathematics, anxiety, and the brain. *Rev. Neurosci.* **2017**, *28*, 417–429. [CrossRef]
41. Tseng, Y.L.; Lu, C.F.; Wu, S.M.; Shimada, S.; Huang, T.; Lu, G.Y. A functional near-infrared spectroscopy study of state anxiety and auditory working memory load. *Front. Hum. Neurosci.* **2018**, *12*, 313. [CrossRef]
42. Thayer, J.F.; Lane, R.D. A model of neurovisceral integration in emotion regulation and dysregulation. *J. Affect. Disord.* **2000**, *61*, 201–216. [CrossRef]
43. Barbas, H.; Saha, S.; Rempel-Clower, N.; Ghashghaei, T. Serial pathways from primate prefrontal cortex to autonomic areas may influence emotional expression. *BMC Neurosci.* **2003**, *4*, 25. [CrossRef] [PubMed]
44. Brugnera, A.; Zarbo, C.; Adorni, R.; Tasca, G.A.; Rabboni, M.; Bondi, E.; Compare, A.; Sakatani, K. Cortical and cardiovascular responses to acute stressors and their relations with psychological distress. *Int. J. Psychophysiol.* **2017**, *114*, 38–46. [CrossRef] [PubMed]
45. Hembree, R. The nature, effects, and relief of mathematics anxiety. *J. Affect. Disord.* **2000**, *61*, 201–216. [CrossRef]

Article

Asymmetry of Cerebellar Lobular Development in Ferrets

Kazuhiko Sawada [1,*], Shiori Kamiya [1] and Ichio Aoki [2,3]

[1] Department of Nutrition, Faculty of Medical and Health Sciences, Tsukuba International University, Tsuchiura, Ibaraki 300-0051, Japan; s-kamiya@tius.ac.jp

[2] Department of Molecular Imaging and Theranostics, National Institute of Radiological Sciences, National Institutes for Quantum and Radiological Science and Technology (QST), Chiba 263-8555, Japan; aoki.ichio@qst.go.jp

[3] Institute for Quantum Life Science (iQLS), QST, Chiba 263-8555, Japan

* Correspondence: k-sawada@tius.ac.jp; Tel.: +81-29-883-6032

Received: 23 March 2020; Accepted: 4 April 2020; Published: 5 May 2020

Abstract: The ferret cerebellum is anteriorly right-lateralized and posteriorly left-lateralized. This study characterized the left/right difference in ferret cerebellar lobular morphology using 3D-rendered magnetic resonance images of fixed brains from seven male and seven female ferrets on postnatal day 90. Asymmetrical lobular morphology showed asymmetrical sublobular development in the anterior vermis, lobulus simplex, and ansiform lobules and additional grooves asymmetrically appearing in the paramedian lobule, lobule VI, and ansiform lobules. Although we observed these asymmetric hallmarks in four cerebellar transverse domains in both sexes, there was no left/right difference in their incidence in each domain. Males showed a significantly higher incidence of the additional grooves in the left side of the ansiform lobules than in females. Data were combined and classified as per the asymmetry quotient (AQ) into left- (AQ < 0) and right-dominant (AQ > 0) groups. There were significantly higher incidences of poor sublobular development of ansiform lobules and additional groove appearing in lobule VI on the right than on the left in the left-dominant group. Asymmetric hallmarks visible on the cerebellar surface of ferrets are relevant to the left-biased volume asymmetry of the central zone of cerebellar transversus domains containing lobule VI and ansiform lobules.

Keywords: asymmetry; sex difference; MRI; volumetry; cerebellum; ferret

1. Introduction

Studies have reported cerebellar volume laterality in humans [1], nonhuman primates [2,3], and carnivores [4], including ferrets [5,6]. Interestingly, regional differences in cerebellar volume laterality revealed distinctive torque morphology formed clockwise (anteriorly left-biased and posteriorly right-biased) in humans [2,7] and counterclockwise (anteriorly right-biased and posteriorly left-biased) in ferrets [5,6]. The posterior cerebellum, including lobule VI and ansiform lobules, is functionally lateralized, that is, the right-side controls language and working memory [8,9], while the left side controls cognition [10,11]. Such functional asymmetry is involved in characteristic functional cerebellar organization via cerebrum–cerebellum connections linked contralaterally with the association cortex, not the motor cortex [9]. Thus, cerebellar asymmetry is evaluated volumetrically and functionally but is not examined on the basis of detailed cerebellar morphology. This study characterized the left/right difference in the cerebellar lobular morphology in male and female ferrets using 3D-rendered images obtained from *ex vivo* magnetic resonance (MR) images.

2. Materials and Methods

2.1. Samples

Anatomical MR images of fixed brains from seven male and seven female ferrets on postnatal day 90 obtained in our previous studies [5,6] were used in this study. A preclinical 7.0 T magnetic resonance imaging (MRI) system with a 400 mm inner-diameter-bore magnet (Kobelco and Jastec, Kobe, Japan) and an AVANCE-I console (Bruker BioSpin, Ettlingen, Germany) acquired 3D MR images covering the entire fixed brains using the rapid acquisition with relaxation enhancement (RARE) sequence with the following parameters: Repetition time (TR) = 300 ms; echo time (TE) = 9.6 ms (effective TE = 19.2 ms); RARE factor = 4; field of view (FOV) = 32 × 32 × 40 mm^3; acquisition matrix = 256 × 256 × 256; voxel size = 125 × 125 × 156 μm^3; number of acquisitions (NEX) = 2; and total scan time = 2 h 43 min 50 s.

2.2. Volumetry

Volumetric analysis was conducted using 3D MR images of the cerebellum. The left and right cerebellar sides were divided at the midline, which was defined by the position of the cerebral longitudinal fissure. Cerebellar transverse domains were segmented semi-automatically into four zones: Anterior zone (AZ; lobules I–V of the vermis), central zone anterior (CZa; lobules VI and the lobules simplex), central zone posterior (CZp; lobules VII of the vermis and ansiform lobules), and posterior zone (PZ; lobules VIII–IXa of the vermis and paramedian lobule) [5]. Segmented areas of each domain were measured using the SliceOmatic software version 4.3 (TomoVision, Montreal, Canada), and the volume of each domain was calculated by multiplying the combined areas by the slice thickness (156.25 μm). The asymmetry quotient (AQ) was calculated using the formula $((R - L)/\{(R + L) \times 0.5\})$ and was used to assess the volume laterality of each domain. The direction of asymmetry was indicated by the AQ values: Positive value = rightward bias and negative value = leftward bias [12].

2.3. 3D-Rendered Images

The cerebellar cortex was segmented semi-automatically on MR images using Amira ver. 5.2 (Visage Imaging, Inc., San Diego, CA, USA) on the basis of image contrast. The images were rendered in 3D using the surface projection algorithm, which best visualizes the surface. Then, the 3D-rendered images were rotated and manipulated to best visualize the cerebellar morphology by linear registration using SliceOmatic software. Finally, asymmetric morphological changes (i.e., poor sublobular development and additional grooves) were noted in the left and right sides of each cerebellar transverse domain on the 3D-rendered images. The cerebellar lobules/sublobules were identified with reference to the textbook of Lawes and Andrews (1998) [13].

2.4. Statistical Analysis

The incidence of asymmetry in sublobular development and additional grooves was estimated in each cerebellar transverse domain segmented with reference to expression patterns of zebrin II/aldolase C in other mammals [14,15]. Left-/right-side or sexual differences were evaluated using the chi-square test. In addition, the relationship between morphological asymmetry and volume laterality in each cerebellar transverse domain was evaluated. Then, data from males and females were combined and classified on the basis of AQ values into left-dominant ($AQ < 0$) and right-dominant ($AQ > 0$) groups. The incidence of asymmetry in sublobular development and additional grooves was estimated in left- and right-dominant groups of all transverse domains, and left-/right-side or intergroup differences were evaluated using the chi-square test.

2.5. Ethics

All experiments were conducted in accordance with the Guide for the Care and Use of Laboratory Animals by the National Institutes of Health (NIH, Bethesda, MD, USA). The study, and all its procedures, was approved by the Institutional Animal Care and Use Committee of Tsukuba International University, Japan. All efforts were made to minimize the number of animals used and their suffering.

3. Results

3.1. Types of Cerebellar Gross Anatomical Asymmetry

The cerebellar morphological asymmetry was roughly classified into two types: Asymmetrical lobular/sublobular development and asymmetrical appearance of additional grooves. Figure 1 illustrates the types of morphological asymmetry of the ferret cerebellum.

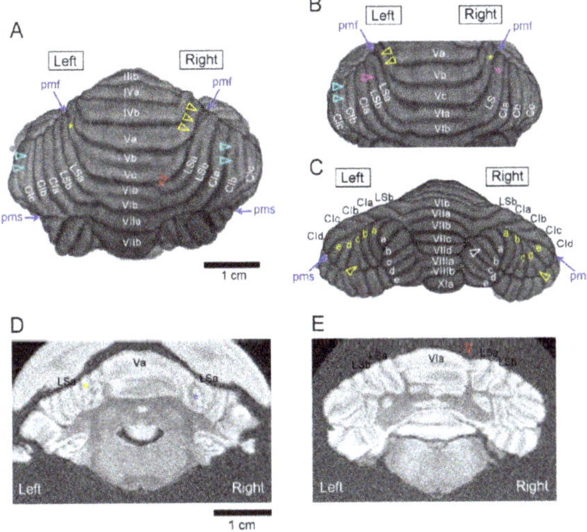

Figure 1. Typical asymmetric morphology in the ferret cerebellum. (**A**) Dorsal view of 3D-rendered cerebellar image showing asymmetrically poor development of the sublobule in the anterior vermis and additional grooves emerging asymmetrically in the lobulus simplex (LS) and ansiform lobules. Sublobule Vb/c was visible at the left end on the cerebellar surface (yellow asterisk) but not at the right end, which was covered by adjacent sublobules (yarrow arrowheads). Additional grooves were found asymmetrically in sublobule VIa of the right paravermis (red arrowhead) and crus I sublobules (right CIb and left CIc) (light-blue arrowheads). (**B**) Dorsal view of 3D-rendered cerebellar image showing asymmetrically poor development of the LS. The sulcus demarcating LS sublobules was present on the left (purple arrowhead) but not on the right (purple asterisk). Sublobule Vc was visible at the right end on the cerebellar surface (yellow asterisk) but not at the left end, which was covered by adjacent sublobules (yarrow arrowheads). Additional grooves emerging asymmetrically were seen in the left CIb (light-blue arrowheads). (**C**) Posterior view of 3D-rendered cerebellar image showing additional grooves emerging asymmetrically in crus II (CII) sublobules (left CIIc and right CIId) (yellow arrowheads) and the paramedian sublobule (right PMb) (white arrowhead). Yellow a–e letters indicate CII sublobules; white a–e letters indicate PM sublobules. (**D**) Coronal (transaxial) MR image showing asymmetrically poor sublobular development in the anterior vermis. The blue asterisk indicates the poorly developed right Vc sublobule, which was covered by the adjacent sublobules Va and LSa on the cerebellar surface. The yellow asterisk indicates contralateral Vc sublobules, which were visible on the cerebellar surface. (**E**) Coronal (transaxial) MR image showing an additional groove appearing in sublobule VI of the right paravermis (red arrowhead). CI, crus I of ansiform lobule; CII, crus II of ansiform lobule; LS, lobulus simplex; PM, paramedian lobule; MR, magnetic resonance.

Sublobules were poorly developed on the left or right, which was covered by adjacent lobules/sublobules (Figure 1A,D). Such asymmetry was seen primarily in the anterior vermis and secondarily in the lobulus simplex and ansiform lobules. The asymmetrical lobular development was also characterized by underdeveloped sulci demarcating the sublobules on the left or right (Figure 1B). This asymmetry was found primarily in lobule VI and the lobulus simplex. Additional grooves appeared asymmetrically in sublobules of paravermian lobule VI, ansiform lobules, and the paramedian lobule (Figure 1A,C). Particularly, the additional groove appearing in the right paravermian lobule VI was distinct from those appearing in other lobules; it infolded loosely and made a distinctive asymmetric profile in lobule VI on coronal MR images (Figure 1E).

3.2. *Incidence of Cerebellar Asymmetrical Morphology*

The highest incidence of asymmetrically poor development of sublobules was observed in the AZ (corresponding to the anterior vermis) on both left and right sides in 57.1% of males and on the right in 57.1% of females. The incidence did not differ between left and right sides and at ipsilateral sides between sexes (Table 1). We also observed no additional grooves asymmetrically appearing in AZ sublobules (Table 2).

Table 1. Incidences of asymmetrically poor sublobule development in male and female ferrets.

Transverse Domains	Left Side		Right Side	
Anterior zone				
Male	57.1%	(4/7)	57.1%	(4/7)
Female	14.3%	(1/7)	57.1%	(4/7)
Central zone anterior				
Male	0%	(0/7)	42.9%	(3/7)
Female	42.9%	(3/7)	0%	(0/7)
Central zone posterior				
Male	0%	(0/7)	28.6%	(2/7)
Female	0%	(0/7)	42.9%	(3/7)
Posterior zone				
Male	0%	(0/7)	0%	(0/7)
Female	0%	(0/7)	0%	(0/7)

The number of individuals exhibiting the poor sublobular development at each side is shown in parentheses.

Table 2. Incidences of asymmetrically appearing additional cerebellum indentations in male and female ferrets.

Transverse Domains	Left Side		Right Side	
Anterior zone				
Male	0%	(0/7)	0%	(0/7)
Female	0%	(0/7)	0%	(0/7)
Central zone anterior				
Male	0%	(0/7)	42.9%	(3/7)
Female	0%	(0/7)	42.9%	(3/7)
Central zone posterior				
Male	57.1%	(4/7) *	28.6%	(2/7)
Female	0%	(0/7)	0%	(0/7)
Posterior zone				
Male	42.9%	(3/7)	57.1%	(4/7)
Female	28.6%	(2/7)	28.6%	(2/7)

The number of individuals showing the additional grooves at each side is shown in parentheses. * $p < 0.05$ male vs. female (chi-square test).

In the CZa, the lobulus simplex was poorly developed on the right in 42.9% of males and on the left in 42.9% of females (Table 1). Additional grooves appeared in paravermian lobule VI on the right in 42.9% of males and 28.6% of females but not on the left in both sexes (Table 2).

In the CZp, sublobules in ansiform lobules were poorly developed on the right in 28.6% of males and 42.9% of females. Although ansiform sublobules on the left were well developed in both males and females, we did not detect a statistical left/right difference in the incidence of asymmetrically poor development of ansiform sublobules in both sexes (Table 1). Additional grooves were found in ansiform lobules on the left in 57.1% of males and on the right in 28.6% of males (Table 2). However, there were no additional grooves in either the left or the right of ansiform lobules in females (Table 2). We observed a significant sexual difference in the incidence of additional grooves only in the left CZp, which was significantly higher in males compared to females (Table 2).

There were no asymmetrically poorly developed sublobules in the PZ in both males and females (Table 1). Additional grooves appeared asymmetrically on the left and right of the paramedian sublobules in both males and females without a significant sexual difference and a left/right difference (Table 2).

3.3. Relationship with Regional Cerebellar Volume Laterality

Volumes of left and right sides and mean AQ values of each transverse domain are shown in Table S1 in the Supplementary Materials.

We observed significant differences in the incidence of asymmetry in sublobular development and additional grooves appearing in the left-dominant group between the left and right sides of the CZa and CZp. Poor sublobular development appeared more frequently in the right CZp compared to the left CZp in the left-dominant group (Table 3). There was a significantly greater incidence of additional grooves in the right CZa compared to the left CZa in the left-dominant group (Table 4). Therefore, the right-dominant appearance of the poor development of CZp sublobules and additional grooves in the CZa might be associated with left-lateralized volumes of these cerebellar domains.

Table 3. Incidence of asymmetrically poor development of cerebellar sublobules in ferrets with left- or right-dominant volume laterality in cerebellar transverse domains.

Transverse Domains	Left-Dominant (AQ < 0)		Right-Dominant (AQ > 0)	
Anterior zone				
Left side	66.7%	(2/3)	27.3%	(3/11)
Right side	33.3%	(1/3)	63.6%	(7/11)
Central zone anterior				
Left side	30.0%	(3/10)	50.0%	(2/4)
Right side	20.0%	(2/10)	25.0%	(1/4)
Central zone posterior				
Left side	0%	(0/10)	0%	(0/4)
Right side	40.0%	(4/10) *	25.0%	(1/4)
Posterior zone				
Left side	0%	(3/8)	0%	(0/6)
Right side	0%	(4/8)	0%	(0/6)

The number of individuals exhibiting the poor sublobular development at each side is shown in parentheses.
* $p < 0.05$ left side vs. right side in left-dominant group (chi-square test).

Table 4. Incidence of asymmetrically appearing additional grooves in ferrets with left- or right-dominant volume laterality in cerebellar transverse domains.

Transverse Domains	Left-Dominant (AQ < 0)		Right-Dominant (AQ > 0)	
Anterior zone				
Left side	0%	(0/3)	0%	(0/11)
Right side	0%	(0/3)	0%	(0/11)
Central zone anterior				
Left side	0%	(0/10)	0%	(0/4)
Right side	40.0%	(4/10) *	50.0%	(2/4)
Central zone posterior				
Left side	30.0%	(3/10)	25.0%	(1/4)
Right side	20.0%	(2/10)	0%	(0/4)
Posterior zone				
Left side	37.5%	(3/8)	33.3%	(2/6)
Right side	50.0%	(4/8)	33.3%	(2/6)

The number of individuals showing the additional grooves at each side is shown in parentheses. * $p < 0.05$ left side vs. right side in left-dominant group (chi-square test).

4. Discussion

The cerebellum is functionally lateralized by making cerebrum–cerebellum connections linked contralaterally with the association cortex, not the motor cortex [9,16]. Right-handed humans showed mirror linkage of the torque asymmetry between the cerebrum and cerebellum, i.e., counterclockwise versus clockwise [7]. The left-lateralized volume of the CZp in the ferret cerebellum plays a role in poor sublobular development in right ansiform lobules. Such left-lateralized development might be involved in left CZp function, such as place-based navigation [10].

Another asymmetric profile appearing in relation to the cerebellar volume laterality is the additional groove in the right paravermian lobule VI, especially in ferrets with the left-biased volume of the CZa. A meta-analysis of task-based neuroimaging studies has shown that the left lobule VI is involved in spatial processing, while the right lobule VI is involved in language and working memory processing [8]. As cerebellar morphological lateralization is related to cerebrum–cerebellum connections [9], the right lobule VI might infold by an enchantment of connectivity to the left cerebral hemisphere.

There is a significant male-over-female incidence in additional grooves appearing in the left CZp, which mainly emerge in left ansiform lobules, indicating sexual differences in CZp function. Sex-related paw preferences have a significant correlation with cerebellar volume asymmetry in dogs [4]. However, ansiform lobules might not be related to paw preferences. In humans, handedness is not related to the superior posterior cerebellar lobule volume, including crura I and II, in terms of interhemispheric connectivity in cortical systems [17]. In contrast, the emergence of additional indentations, called secondary and tertiary sulci, in the cerebral cortex might be associated with regional cortical growth [18]. A preliminary study revealed the possibility of adult cerebellar neurogenesis in ferrets [19], as reported in humans and mice [20]. As volumetric asymmetry of the ferret cerebellum is evident during puberty to adolescence [6], regional cerebellar growth by adult neurogenesis might be involved in the appearance of additional sex-related grooves. Further studies are needed to elucidate this.

5. Conclusions

A distinctive cerebellar morphological asymmetry has been documented in humans [7], nonhuman primates [2], and carnivores [4,5], although the direction and degree of asymmetry are altered by species and/or sexes [2–7] but not by genetic factors [7]. As the rodent cerebellum developed symmetrically [21], ferrets would be useful models for investigating the pathogenesis underlying human neurodevelopmental disorders with disturbances in morphological cerebellar asymmetry, such as autistic spectrum disorder [22], schizophrenia [23,24], dyslexia [25], and attention-deficit hyperactivity disorder [26]. Asymmetric hallmarks of the cortical surface morphology visible on the cerebellar surface of ferrets are relevant to the left-biased volume asymmetry of the CZ of cerebellar transversus domains containing lobule VI and ansiform lobules in ferrets. The findings provide a new insight into assessments of cerebellar asymmetry on the basis of cerebellar surface morphology.

Supplementary Materials: The following is available online at http://www.mdpi.com/2073-8994/12/5/735/s1, Table S1: Volume and asymmetry quotient of cerebellar cortex in male and female ferrets.

Author Contributions: K.S. designed the study; K.S. and S.K. performed the experiments and analyzed the data; K.S. and I.A. wrote the manuscript; the final version of the manuscript was approved by all authors. All authors have read and agreed to the published version of the manuscript.

Funding: This study was supported by Grant from Tsukuba International University, and partly supported by the Center of Innovation Program (Japan Science and Technology Agency; JST) for MRI devices and AMED under Grant Number 17dm0107066h for imaging technologies.

Acknowledgments: The authors wish to thank Nobuhiro Nitta (Molecular Imaging Center, National Institute of Radiological Sciences, Chiba, Japan) for the MRI measurements. The authors would like to thank Enago (www.enago.jp) for the English language review.

Conflicts of Interest: The authors declare no conflict of interest.

References

1. Synder, P.J.; Bilder, R.M.; Wu, H.; Bogerts, B.; Lieberman, J.A. Cerebellar volume asymmetries are related to handedness: A quantitative MRI study. *Neuropsychologia* **1995**, *33*, 407–419. [CrossRef]
2. Phillips, K.; Hopkins, W.D. Exploring the relationship between cerebellar asymmetry and handedness in chimpanzees (*Pan troglodytes*) and capuchins (*Cebus apella*). *Neuropsychologia* **2007**, *45*, 2333–2339. [CrossRef] [PubMed]
3. Cantalupo, C.; Freeman, H.; Rodes, W.; Hopkins, W. Handedness for tool use correlates with cerebellar asymmetries in chimpanzees (*Pan troglodytes*). *Behav. Neurosci.* **2008**, *122*, 191–198. [CrossRef] [PubMed]
4. Koyun, N.; Aydinlioğlu, A.; Aslan, K. A morphometric study on dog cerebellum. *Neurol. Res.* **2011**, *33*, 220–224. [CrossRef]
5. Sawada, K.; Horiuchi-Hirose, M.; Saito, S.; Aoki, I. Male prevalent enhancement of leftward asymmetric development of the cerebellar cortex in ferrets (*Mustela putorius*). *Laterality* **2015**, *20*, 723–737. [CrossRef]
6. Sawada, K.; Aoki, I. Age-dependent sexually-dimorphic asymmetric development of the ferret cerebellar cortex. *Symmetry* **2017**, *9*, 40. [CrossRef]
7. Rosch, R.E.; Ronan, L.; Cherkas, L.; Gurd, J.M. Cerebellar asymmetry in a pair of monozygotic handedness-discordant twins. *J. Anat.* **2010**, *217*, 38–47. [CrossRef]
8. Valera, E.M.; Faraone, S.V.; Biederman, J.; Poldrack, R.A.; Seidman, L.J. Functional neuroanatomy of working memory in adults with attention-deficit/hyperactivity disorder. *Biol. Psychiatry* **2005**, *57*, 439–447. [CrossRef]
9. Wang, D.; Buckner, R.L.; Liu, H. Cerebellar asymmetry and its relation to cerebral asymmetry estimated by intrinsic functional connectivity. *J. Neurophysiol.* **2013**, *109*, 46–57. [CrossRef]
10. Iglói, K.; Doeller, C.F.; Paradis, A.L.; Benchenane, K.; Berthoz, A.; Burgess, N.; Rondi-Reig, L. Interaction between hippocampus and cerebellum crus I in sequence-based but not place-based navigation. *Cereb. Cortex* **2015**, *25*, 4146–4154. [CrossRef]
11. Sokolov, A.A. The cerebellum in social cognition. *Front. Cell Neurosci.* **2018**, *12*, 145. [CrossRef]
12. Hopkins, W.D.; Marino, L. Asymmetries in cerebral width in nonhuman primate brains as revealed by magnetic resonance imaging (MRI). *Neuropsychologia* **2000**, *38*, 493–499. [CrossRef]
13. Lawes, I.N.C.; Andrews, P.L.R. Neuroanatomy of the ferret brain. In *Biology and Diseases of the Ferret*, 2nd ed.; Fox, Ed.; Lippincott Williams and Wilkins: Baltimore, MD, USA, 1998; pp. 71–102.
14. Ozol, K.; Hayden, J.M.; Oberdick, J.; Hawkes, R. Transverse zones in the vermis of the mouse cerebellum. *J. Comp. Neurol.* **1999**, *412*, 95–111. [CrossRef]
15. Sillitoe, R.V.; Hawkes, R. Whole-mount immunohistochemistry: A high-throughput screen for patterning defects in the mouse cerebellum. *J. Histochem. Cytochem.* **2002**, *50*, 235–244. [CrossRef]
16. Siniscalchi, M.; Franchini, D.; Pepe, A.M.; Sasso, R.; Dimatteo, S.; Vallortigara, G.; Quaranta, A. Volumetric assessment of cerebral asymmetries in dogs. *Laterality* **2011**, *16*, 528–536. [CrossRef]
17. Rosch, R.E.; Cowell, P.E.; Gurd, J.M. Cerebellar asymmetry and cortical connectivity in monozygotic twins with discordant handedness. *Cerebellum* **2018**, *17*, 191–203. [CrossRef]

18. Sawada, K.; Fukunishi, K.; Kashima, M.; Imai, N.; Saito, S.; Aoki, I.; Fukui, Y. Regional difference in sulcal infolding progression correlated with cerebral cortical expansion in cynomolgus monkey fetuses. *Congenit. Anom. (Kyoto)* **2017**, *57*, 114–117. [CrossRef]
19. Kamiya, S.; Sawada, K. Immunohistochemical characterization of postnatal changes in cerebellar cortical cytoarchitectures in ferrets. *Anat. Rec. (Hoboken)*, in press.
20. Ahlfeld, J.; Filser, S.; Schmidt, F.; Wefers, A.K.; Merk, D.J.; Glaß, R.; Herms, J.; Schüller, U. Neurogenesis from Sox2 expressing cells in the adult cerebellar cortex. *Sci. Rep.* **2017**, *7*, 6137. [CrossRef]
21. Dorr, A.E.; Lerch, J.P.; Spring, S.; Kabani, N.; Henkelman, R.M. High resolution three-dimensional brain atlas using an average magnetic resonance image of 40 adult C57Bl/6J mice. *Neuroimage* **2008**, *42*, 60–69. [CrossRef]
22. Hodge, S.M.; Makris, N.; Kennedy, D.N.; Caviness, V.S., Jr.; Howard, J.; McGrath, L.; Steele, S.; Frazier, J.A.; Tager-Flusberg, H.; Harris, G.J. Cerebellum, language, and cognition in autism and specific language impairment. *J. Autism Dev. Disord.* **2010**, *40*, 300–316. [CrossRef] [PubMed]
23. Levitt, J.J.; McCarley, R.W.; Nestor, P.G.; Petrescu, C.; Donnino, R.; Hirayasu, Y.; Kikinis, R.; Jolesz, F.A.; Shenton, M.E. Quantitative volumetric MRI study of the cerebellum and vermis in schizophrenia: Clinical and cognitive correlates. *Am. J. Psychiatry* **1999**, *156*, 1105–1107. [PubMed]
24. Sheng, J.; Zhu, Y.; Lu, Z.; Liu, N.; Huang, N.; Zhang, Z.; Tan, L.; Li, C.; Yu, X. Altered volume and lateralization of language-related regions in first-episode schizophrenia. *Schizophr. Res.* **2013**, *148*, 168–174. [CrossRef] [PubMed]
25. Kibby, M.Y.; Fancher, J.B.; Markanen, R.; Hynd, G.W. A quantitative magnetic resonance imaging analysis of the cerebellar deficit hypothesis of dyslexia. *J. Child Neurol.* **2008**, *23*, 368–380. [CrossRef] [PubMed]
26. Castellanos, F.X.; Giedd, J.N.; Marsh, W.L.; Hamburger, S.D.; Vaituzis, A.C.; Dickstein, D.P.; Sarfatti, S.E.; Vauss, Y.C.; Snell, J.W.; Lange, N.; et al. Quantitative brain magnetic resonance imaging in attention-deficit hyperactivity disorder. *Arch. Gen. Psychiatry* **1996**, *53*, 607–616. [CrossRef] [PubMed]

Article

Does Brain Lateralization Affect the Performance in Binary Choice Tasks? A Study in the Animal Model *Danio rerio*

Maria Elena Miletto Petrazzini [1,*], Alessandra Pecunioso [2], Marco Dadda [2] and Christian Agrillo [2,3]

1 Department of Biomedical Sciences, University of Padova, Via Bassi 58-b, 35131 Padova, Italy
2 Department of General Psychology, University of Padova, Via Venezia 8, 35131 Padova, Italy; alessandra.pecunioso@studenti.unipd.it (A.P.); marco.dadda@unipd.it (M.D.); christian.agrillo@unipd.it (C.A.)
3 Padua Neuroscience Center, University of Padova, Via 8 Febbraio 2, 35122 Padova, Italy
* Correspondence: m.e.milettopetrazzini@qmul.ac.uk or mariaelena.milettopetrazzini@gmail.com

Received: 30 May 2020; Accepted: 31 July 2020; Published: 4 August 2020

Abstract: Researchers in behavioral neuroscience commonly observe the behavior of animal subjects in the presence of two alternative stimuli. However, this type of binary choice introduces a potential confound related to side biases. Understanding whether subjects exhibit this bias, and the origin of it (pre-existent or acquired throughout the experimental sessions), is particularly important to interpreting the results. Here, we tested the hypothesis according to which brain lateralization may influence the emergence of side biases in a well-known model of neuroscience, the zebrafish. As a measure of lateralization, individuals were observed in their spontaneous tendencies to monitor a potential predator with either the left or the right eye. Subjects also underwent an operant conditioning task requiring discrimination between two colors placed on the left–right axis. Although the low performance exhibited in the operant conditioning task prevents firm conclusions from being drawn, a positive correlation was found between the direction of lateralization and the tendency to select the stimulus presented on one specific side (e.g., right). The choice for this preferred side did not change throughout the experimental sessions, meaning that this side bias was not the result of the prolonged training. Overall, our study calls for a wider investigation of pre-existing lateralization biases in animal models to set up methodological counterstrategies to test individuals that do not properly work in a binary choice task with stimuli arranged on the left–right axis.

Keywords: lateralization; side bias; fish; methodological artefacts; symmetry

1. Introduction

In the last three decades, an increasing number of studies have been reported in the field of behavioral neuroscience that are based on animal models. The observation of spontaneous behavior assessed through the use of free choice tests [1,2] or the study of learned abilities as a result of operant conditioning studies [3–5] shed light on the perceptual–cognitive mechanisms of vertebrates. For both methodological approaches, animals typically undergo binary choices with stimuli presented on the left–right axis [6–9]. These are usually either two biologically relevant stimuli (e.g., pieces of food or conspecifics, in the case of free choice tests [7,10,11]) or two sets of neutral stimuli (e.g., two-dimensional figures or pictures, in the case of operant conditioning studies [8,9,12]). In both types of studies, it is not rare to find subjects that repeatedly select the stimulus presented on one specific side (e.g., always the right one). This effect is referred to as side bias or side preference [13,14], which is the tendency to select one alternative stimulus over others even though there is not any identified reason, and belongs to the category of idiosyncratic choice biases [15]. Such biases have also been widely observed in human research (e.g., [15]) and may represent a potential confounding factor that affects the conclusions of an experiment. For instance, Andrade et al. [16] found that rats commonly show left or right bias in spontaneous arm preference in a T-maze, a fact that might have impacted the conclusions of several studies on spatial abilities, learning, and memory in rats tested with this type of apparatus. Understanding the reason for a consistent preference for a stimulus placed on the same side of the apparatus is fundamental. A random choice between two stimuli, whose presentation is counterbalanced across trials between left and right, might be due to two opposite reasons: subjects might not have the cognitive abilities to discriminate between the two stimuli, or they might display such cognitive skills but have side biases that prevent the experimenters from testing their hypotheses. Dissociating between these two alternative explanations is necessary for interpreting subjects' performance.

Side biases might have at least two different origins. First, they might be the result of the experimental context. For instance, due to a repeated number of stimulus presentations used in operant conditioning studies, animals might have started to develop a preference for one side of the apparatus (e.g., [17,18]) because this strategy permits obtaining food rewards approximately 50% of the time (avoiding any need to understand which is the discriminative cue between two stimuli). Second, side biases may be present at the onset of the training because of an intrinsic side preference of the subjects [19–23]. Recently, it has been suggested that, at least in human participants, this type of bias might emerge regardless of the peculiarity of the experimental context or procedure, or any kind of pre-existing side preferences of the individuals. By investigating computational models of decision-making, Lebovich et al. [15] found that idiosyncratic choice biases may emerge naturally from the neural circuits underlying decision-making. In this sense, they would represent a virtually inevitable side effect of all binary choice tasks. A recent study requiring rhesus monkeys to choose between two colors by making saccadic movements [24] suggested that when the reward is unpredictable (for instance, when food items are randomly assigned to the two colors), animals may start to exhibit sub-optimal exploratory behaviors, which would be the result of very slow learning, or reward memory, in the attempt to face the unpredictability of non-stationary environments.

Side biases related to the experimental context or procedure can be partially adjusted, for instance, by adopting a correction procedure, such as presenting the target stimulus on the non-preferred side of the subject, to correct their acquired tendency to turn in one direction. On the contrary, side biases related to pre-existing preferences of subjects might be more difficult to modify if stimuli are exclusively arranged on the left–right axis. Therefore, it becomes fundamental to discriminate the type of side bias (spontaneous vs. acquired) observed in a binary choice task.

Intrinsic side bias is believed to be the result of brain lateralization. Functional brain asymmetries were initially thought to be a human peculiarity, but evidence collected in the last few decades has shown that the asymmetries are a widespread principle of nervous system organization. Functional right–left brain asymmetries have been found in a wide range of species [25–31]. For instance, when exploring a predator, some fish species exhibit a preference for monitoring it with a specific eye [32], as a result of a brain specialization that allows them to encode predators' features with one particular side of the brain (the one contralateral to the eye used to inspect the predator). The possibility exists that strongly lateralized individuals are more susceptible to side biases in binary choices than poorly lateralized ones [33].

In the last few decades, the teleost fish *Danio rerio*—also known as zebrafish—has become one of the most adopted models in genetics, neuroscience, and developmental biology [34,35]. Despite the wide range of neurobiological investigations, the cognitive skills of this species remain largely unknown. Only recently, the methodology for the study of cognition in zebrafish has been improved, a fact that permitted the discovery of the existence of numerical [36,37], spatial [38,39], and perceptual [40,41] abilities in this species. Zebrafish also exhibit a wide range of lateralized behaviors, including conspecifics inspection [42], escape response [43], and foraging behavior [44]. However, no study has directly focused on the relation between brain lateralization and side bias in binary choices of this important animal model.

Here, we tested the hypothesis according to which lateralized individuals are more likely to exhibit side biases than poorly lateralized ones in a binary choice test. To achieve this goal, we observed zebrafish behavior in both a spontaneous choice task and an operant conditioning task. To assess the degree of lateralization of subjects, we used a test known as a detour test [32,45,46] that sheds light on the spontaneous preference for the use of one eye (and, hence, one side of the brain) in the presence of a potential predator. The same subjects were observed in a color discrimination task. They were required to discriminate between two rectangles differing in color, one placed on the right and one placed on the left, to obtain a food reward. Unlike many other training studies on cold-blooded vertebrates [47,48], here we used an automated procedure to guarantee extensive training comparable to that used with mammals and birds. This permitted us to establish whether side biases are acquired throughout the experimental sessions because of the prolonged presentation of stimuli. We correlated the behavior in the detour test to the choices for one side of the experimental tank of the color discrimination task to see whether there was any significant link between brain lateralization and performance in the operant conditioning task.

2. Methods

2.1. Subjects

We tested 16 adult zebrafish (9 females, with approximate ages of 8–14 months). Fish were maintained at the comparative psychology laboratory of the Department of General Psychology (University of Padova) in mixed-sex groups of 20–30 individuals (150 L in each tank). The aquaria each had a 14:10 h light to dark (L:D) photoperiod and an 18 W fluorescent light. Aquaria were provided with air filters, natural gravel, and live plants at a temperature of 25 ± 1 °C. Fish were fed daily twice: once with commercial food flakes and once with live brine shrimps (*Artemia salina*). Each subject was singly moved into a 50 × 19 × 32 cm tank in the week preceding our experiment. Four smaller conspecifics were inserted in the tank to reduce the possibility that social isolation might have affected the subjects' behavior in the two tasks. The local ethics committee of the University of Padova approved the study (permit number: 37/2016).

Fish were observed in two different tasks (presented according to a pseudorandom order): (a) a detour test to assess the degree of each subject's lateralization, and (b) a color discrimination task to assess their ability to learn a conditioned rule.

2.2. Lateralization Test: Detour Test

In this test, subjects typically swim along a runway until they encounter a dummy predator behind a barrier (Figure 1). Eye preference—and the corresponding left or right hemisphere that preferentially encodes the potential predator—is inferred by the subject's direction when leaving the runway [17,34,35].

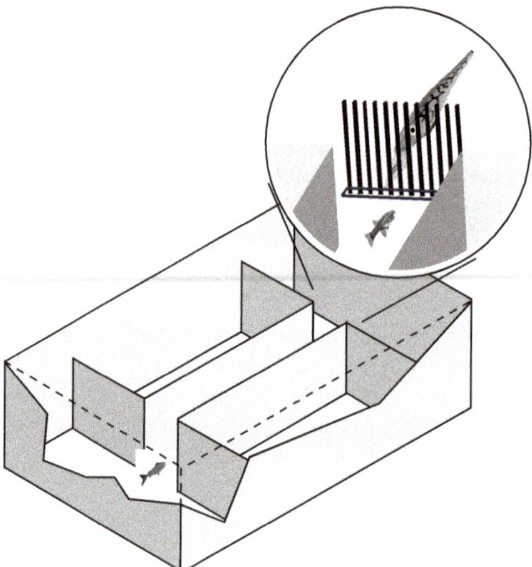

Figure 1. Detour test. A dummy predator was visible behind a barrier at both ends of the runway. The lateralization index was based on 20 observations for each subject.

2.2.1. Stimuli and Apparatus

The apparatus was a modified version used in previous studies of fish [27,33] and consisted of a glass tank (60 × 94 × 36 cm) filled with 10 cm of water. A plastic runway (7 × 40 cm) was placed in the middle, permitting subjects to encounter a barrier. Behind the barrier—a series of vertical yellow cylindrical plastic bars (0.05 cm in diameter, spaced at 0.25 cm intervals) that occupied 17 × 17 cm—a dummy predator (Rapala lure no.18, 18 × 2.5 cm) was visible at both ends of the runway. At the two short sides of the apparatus, we placed two 18 W fluorescent lamps to maximize the visibility of the dummy predators.

2.2.2. Procedure

A single subject was inserted in the middle of the runway. Then, by using a pair of fishnets, the subject was gently pushed toward the starting point. In this way, the subject started to swim along the runway until the dummy predator was visible. We recorded twenty consecutive trials. We used the percentage of left and right turns taken by the subject when leaving the runway as an index of lateralization, based on the following formula: (([number of rightward turns − number of leftward turns]/total number of turns) × 100). This index is a continuous variable and fish with a positive score were classified as subjects with a right detour preference whereas individuals with a negative score were classified as subjects with a left detour preference. In the case of individuals turning rightward half of the time, we classified them as non-lateralized [27,33].

2.3. *Color Discrimination Task*

Color discrimination tasks are commonly used in operant conditioning studies to assess learning abilities [49–51]. Zebrafish were required to reach the positive stimulus (e.g., a green square), instead of the negative one (e.g., a blue square), to obtain a food reward.

2.3.1. Apparatus and Stimuli

We used an automated operant conditioning apparatus developed for zebrafish (Zantiks AD, Figure 2). The whole unit's size included the experimental chamber (22 × 30 × 50 cm, width × length ×

height), a computer, and the software (C++ language). The chamber (14 × 20 × 15 cm) was made up of white plastic walls; the floor was made of transparent plastic. The chamber was filled with 8 cm of water. A computer screen (16 × 10 cm) placed beneath the tank presented the stimuli in correspondence with one of the short walls. A wireless router allowed us to use a laptop to run the experiment and collect data. The integrated computer automatically detected the subject's position through an infrared camera placed above the chamber and an infrared source placed below the chamber. Zantiks scripts of the task are available in the Supplementary Materials.

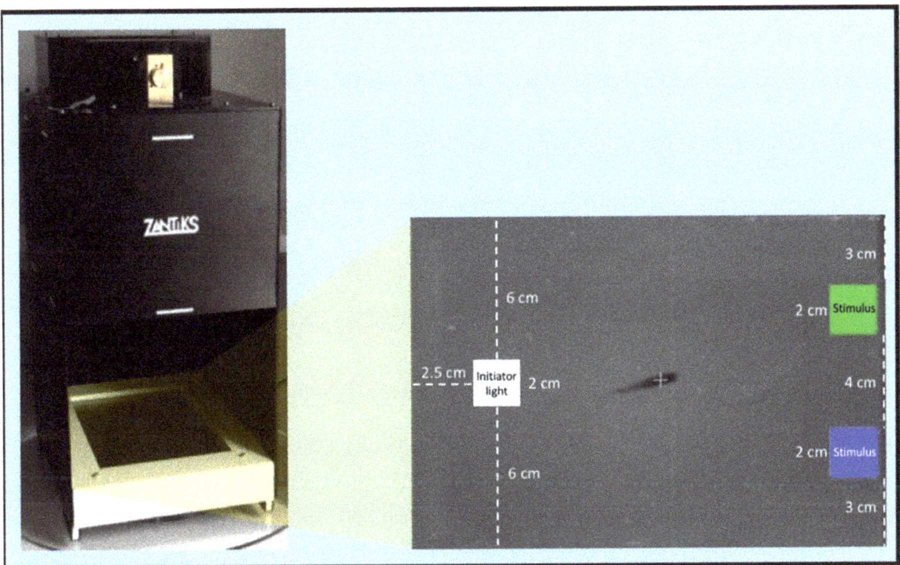

Figure 2. Operant conditioning chamber (Zantiks©) used in the color discrimination task. Stimuli were presented on the bottom of the tank. On the opposite side, we presented two rectangles differing in color (e.g., blue and green). A correct choice triggered the release of a food reward.

Stimuli (2 × 2 cm) consisted of colored rectangles: red (RGB: 255, 0, 0), yellow (RGB: 255, 255, 0), green (RGB: 0, 255, 0), and blue (RGB: 0, 0, 255). An initiator white light rectangle (RGB: 255, 255, 255) appeared at the beginning of each trial.

2.3.2. Procedure

Pre-training: In this stage, we needed to ensure that the zebrafish learned how to trigger the presentation of the stimuli, select one stimulus, and then swim to the feeder and eat the reward. This phase enabled us to eliminate any eventual lack of color discrimination as a general failure to understand the sequence of events composing the trial.

Subjects were moved from their home tanks to the experimental chamber. The pre-training session started after 5 min of habituation and was characterized by 80 overall trials. Each trial started with a white rectangle (2 × 2 cm) that served as an initiator light. This light was presented in correspondence with one short end of the chamber. The initiator light remained visible for a maximum of 10 s; if not selected, the initiator light disappeared for the next 10 s. As soon as a fish approached the initiator light by swimming above the white rectangle, a colored rectangle appeared on the opposite short wall. In the case of the discrimination of yellow vs. red, the colored rectangle was green; in the case of the discrimination of green vs. blue, the colored rectangle was red. The colored rectangle remained visible for a maximum of 10 s. The food reward consisted of a 2 mg portion of commercial flake food, GVG Sera©, and was released above the initiator light every time the subject swam above the colored light.

In the event of no selection of the colored square, the trial was not considered for the analysis and a time-out of 10 s was presented. Subjects completing at least 25 out of 80 trials in a single session started the next step.

Training: Fifty percent of the subjects were trained in yellow and red discrimination (50% with red as a positive stimulus); and 50% of them were trained in blue and green discrimination (50% with blue as a positive stimulus). These specific contrasts were chosen because previous studies showed that zebrafish successfully discriminate between yellow and red, and between blue and green [49].

The experiment was run in a dark environment. As soon as the white initiator light appeared, the trial began. If a fish swam above the light within 10 s, a choice between two different colored squares was given on the opposite side of the tank (see Figure 2 for measures). The interstimulus distance was equal to 4 cm. As the stimuli appeared when subjects triggered the initiator light, placed at 15.5 cm from the stimuli, optical refraction might not have significantly affected the hue, saturation, or brightness of our target colors in any position of the experimental chamber [52]. In the case of the correct choice, we delivered the same food reward used in the pre-training phase in correspondence with the initiator light. On the contrary, if a fish selected the wrong alternative, the two stimuli disappeared, and the bottom of the tank became white until the beginning of next trial. If subjects did not select the initiator light within 10 s, or did not select either stimulus, the stimuli disappeared, and the trial was considered to be an omission. Each training session included 100 trials. The stimulus position on the left–right axis was switched across trials according to a random sequence. The task ended when the fish reached 500 valid trials. (No omissions were included.)

As the data were normally distributed, we used parametric statistics (Kolmogorov–Smirnov on the overall performance of the training, $p = 0.138$).

3. Results

Color discrimination task: a generalized linear mixed model (GLMM) was used to assess whether the performance differed among subjects. A significant inter-individual difference was found ($\chi^2(15) = 529.57$, $p < 0.001$). In particular, binomial tests on the frequency of correct choices showed that 5 subjects out of 16 (31%) selected the color associated with the food reward ($p < 0.05$). No difference in the proportion of choice towards the positive stimulus was found between fish tested in the discrimination between yellow and red and those tested in the discrimination between green and blue (yellow vs. red: 0.424 ± 0.151, blue vs. green: 0.498 ± 0.118; independent *t*-test, $t\ (14) = -1.096$, $p = 0.292$, d = 0.274). Therefore, we pooled the two groups for the subsequent analyses. To measure whether any side bias emerged throughout the training, we calculated the proportion of choices for the stimulus presented on the right side. The proportion of choice for the stimulus on the right side did not change as a function of trials (repeated measures of ANOVA, Greenhouse–Geisser correction, $F(2.515, 37.719) = 1.788$, $p = 0.173$, partial eta squared $\eta_p^2 = 0.106$, Figure 3), meaning that the repeated presentation of stimuli on the left–right axis did not lead to the development of side biases per se.

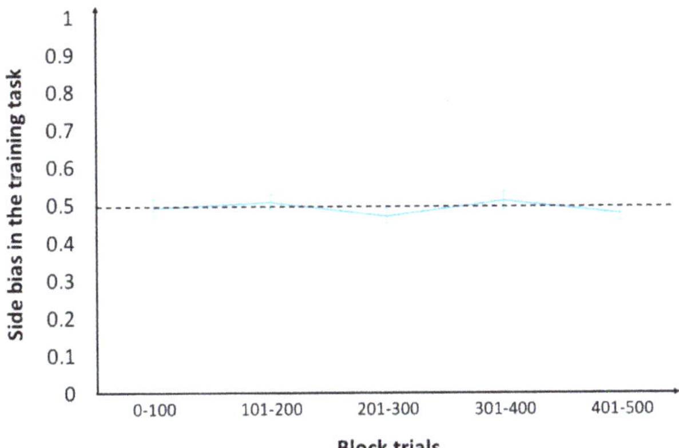

Figure 3. Proportion of choices for the right side of the tank in the color discrimination task. No side bias was developed throughout the 500 trials of the color discrimination task.

Correlation between the detour test and the color discrimination task: a positive correlation was observed between the performance in the detour test and the proportion of choices for the stimulus presented on the right in the color discrimination task (r (14) = 0.921, $p < 0.001$, Bayes factor < 0.001, Figure 4), meaning that the side bias observed in the training task was related to the subject's brain lateralization observed in the detour test. However, no correlation was found between the performance in the detour test and the proportion of correct choices in the training task (r (14) = 0.181, $p = 0.503$, Bayes factor = 4.228, Figure 5), meaning that non-lateralized individuals did not outperform lateralized ones in the present operant conditioning task.

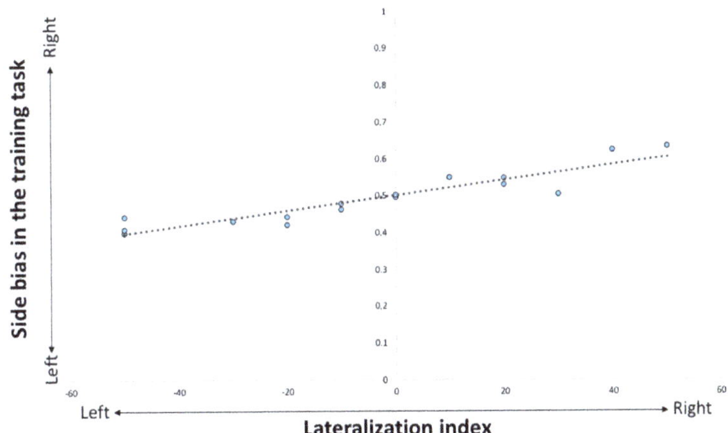

Figure 4. Positive correlation between the lateralization index and the proportion of choices for the right side in the operant conditioning task. Right detour fish tended to select the stimuli of the color discrimination task presented on the right; left detour fish tended to select the opposite stimulus (left).

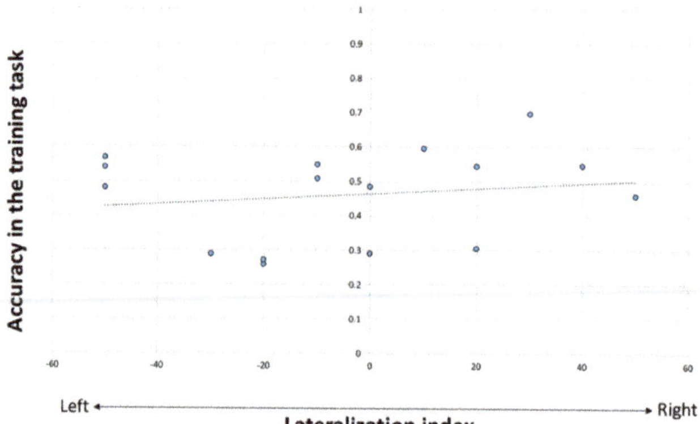

Figure 5. Correlation between the lateralization index and the overall accuracy in the operant conditioning task. Despite a clear side bias of lateralized fish, non-lateralized fish did not statistically outperform lateralized fish.

4. Discussion

We wondered whether brain lateralization impacts the performance of laboratory animals that undergo operant conditioning tasks based on binary choices. To achieve this goal, we used the detour test, one of the most established tests, to assess the direction and degree of brain lateralization, and correlated zebrafish behavior with the performance exhibited in an operant conditioning task requiring selection between one of two colors presented on the left–right axis. We found a strong positive correlation between brain lateralization and the tendency to select repeatedly the stimulus presented on one side of the tank. Our conclusion is supported by two different statistical perspectives: the *p*-value and Bayesian approaches.

Our result has important methodological implications for animal studies based on binary choices. Indeed, most of these studies do not preliminarily assess the degree of hemispheric specialization of their subjects. Therefore, the performance in binary choice tasks may be interpreted because of cognitive abilities or inabilities. We showed that lateralized individuals exhibit this bias from the beginning of the training, a fact that indicates a pre-existing bias rather than an acquired bias developed throughout the trials. Interestingly, we found that zebrafish could not inhibit their tendencies to swim to the preferred side of the experimental chamber even after an extensive training of 500 trials, in which the positive and the negative stimuli were presented randomly on the left–right axis. These subjects may have had the cognitive abilities to solve the color discrimination task, but they might not have been able to perform the correct choice when the correct stimulus was placed on the non-preferred side. This result calls for a cautious interpretation of binary choice tasks, particularly in the case of those studies in which animals are tested only once or twice, such as in free choice tests with subjects observed in their spontaneous preferences for biologically relevant stimuli [53–55]. In this case, it is difficult to establish whether the subjects have a lateralization bias, making data interpretation ambiguous.

The results of our study may be important in behavioral neuroscience because zebrafish are one of the most adopted animal models in genetics, neuroscience, and developmental biology [34,35]. To assess their performance in perceptual–cognitive tasks, many studies used free choice tests [37,43,56] or operant conditioning studies [36,39]. In either case, binary choices are often presented to subjects and the differential performance has been read as the existence of inter-individual variability in the perceptual–cognitive skill under investigation [36,40].

The use of a preliminary test on the brain lateralization of subjects could solve this issue. We suggest two potential solutions. One consists of discarding those subjects who exhibit a robust lateralization bias in the detour test. This would lead to including in the final sample only those fish that could respond to the stimuli. However, this would bring researchers to test a specific subsample of fish: the poorly lateralized ones. There is evidence that strongly and poorly lateralized fish differ for several cognitive factors, including the capacity to perform dual tasks [57], numerical discrimination [58], and spatial tasks [33]. In this sense, testing a sample made by non-lateralized individuals would prevent researchers from generalizing the conclusions to the whole population of zebrafish. Alternatively, detour tests might be useful to identify if there is a large percentage of strongly lateralized individuals. If so, subjects could undergo binary choices with stimuli arranged on the vertical axis (one above and the other below). This experimental strategy, of course, would skip the left–right bias, but could be adopted only after ensuring that the species under investigation does not have a spontaneous preference for selecting a stimulus placed on either the top or bottom of the experimental apparatus (a fact that might represent another side bias that affects the validity of the binary choice tasks).

If strongly lateralized individuals are expected to exhibit strong side bias choice, we should expect non-lateralized individuals to be advantaged in tasks that require learning a discriminative rule between stimuli arranged on the left–right axis. There is indeed evidence that, in some cases, non-lateralized fish outperform lateralized ones; non-lateralized goldbelly topminnows are more accurate when required to bisect a line or to transfer visual information quickly between the two halves of the brain before choosing which shoal to join [33]. We have not found a correlation between the degree of lateralization and the accuracy in the operant conditioning task, but we cannot exclude that an advantage for non-lateralized ones may emerge in other cognitive tasks or other experimental conditions. We must note that a large inter-individual variability in the subjects' performance of the color discrimination task was found, and the overall performance of subjects in our color discrimination task was low. We cannot exclude the possibility that the results of the correlational analyses reported here would not persist in a population of well-trained fish. Developing finer procedures for testing perceptual–cognitive skills in zebrafish is fundamental, given the brain imaging possibilities available in larvae. Such investigation would permit, for instance, mapping behavioral lateralization (such as the one tested here) onto the neural circuits of a zebrafish brain, or to determine neural circuits underlying the emergence of side biases. However, it is worth noting that the subjects' percentage for selecting the positive stimulus at the end of our training (31%) was similar to the percentage of zebrafish reaching the learning criterion in other perceptual–cognitive tasks [36,59]. This suggests that, rather than assuming that our training procedure was largely ineffective in this species, the cognitive skills of this important animal model could barely be shaped by extensive training, compared with other fish species (e.g., guppies [60], redtail splitfins [61], and angelfish [62]).

In conclusion, we found that the behavior exhibited in a test measuring brain lateralization seems to be predictive of the side bias exhibited by zebrafish in binary choices. Lateralized individuals tend to turn to a preferred side of the experimental tank regardless of extensive training, a fact that makes it difficult to understand whether these individuals have the cognitive abilities to solve the task.

Supplementary Materials: The following are available online at http://www.mdpi.com/2073-8994/12/8/1294/s1.

Author Contributions: Conceptualization, M.E.M.P. and C.A.; Data curation, A.P.; Funding acquisition, M.E.M.P.; Investigation, M.D; Methodology, A.P.; Writing—original draft, M.E.M.P., A.P., M.D. and C.A. All authors have read and agreed to the published version of the manuscript.

Funding: This study was supported by 'Stars@unipd' grant (acronym: MetaZeb) from the University of Padova to M.E.M.P.

Acknowledgments: We performed the present work within the scope of the research grant Dipartimenti di Eccellenza entitled "Innovative methods or technologies for assessment, intervention, or enhancement of psychological functions (cognitive, emotional or behavioral)".

Conflicts of Interest: The authors declare no conflict of interest.

References

1. Parrish, A.E.; Beran, M.J. When less is more: Like humans, chimpanzees (*Pan troglodytes*) misperceive food amounts based on plate size. *Anim. Cogn.* **2014**, *17*, 427–434. [CrossRef]

2. Lucon-Xiccato, T.; Miletto Petrazzini, M.E.; Agrillo, C.; Bisazza, A. Guppies discriminate between two quantities of food items but prioritize item size over total amount. *Anim. Behav.* **2015**, *107*, 183–191. [CrossRef]
3. Santacà, M.; Agrillo, C. Perception of the Müller—Lyer illusion in guppies. *Curr. Zool.* **2020**, *66*, 205–213. [CrossRef] [PubMed]
4. Cantlon, J.; Brannon, E. Basic math in monkeys and collegestudents. *PLoS Biol.* **2007**, *5*, e328. [CrossRef] [PubMed]
5. Sovrano, V.A.; Bisazza, A.; Vallortigara, G. Modularity as a fish (*Xenotoca eiseni*) views it: Conjoining geometric and nongeometric information for spatial reorientation. *J. Exp. Psych. Anim. Behav. Proc.* **2003**, *29*, 199–210. [CrossRef] [PubMed]
6. Robins, A.; Rogers, L.J. Complementary and lateralized forms of processing in *Bufo marinus* for novel and familiar prey. *Neurobiol. Learn. Mem.* **2006**, *86*, 214–227. [CrossRef] [PubMed]
7. Giljov, A.N.; Karenina, K.A.; Malashichev, Y.B. An eye for a worm: Lateralisation of feeding behaviour in aquatic anamniotes. *Laterality* **2009**, *14*, 273–286. [CrossRef]
8. Peirce, J.W.; Leigh, A.E.; Kendrick, K.M. Configurational coding, familiarity and the right hemisphere advantage for face recognition in sheep. *Neuropsychologia* **2000**, *38*, 475–483. [CrossRef]
9. Bortot, M.; Agrillo, C.; Avarguès-Weber, A.; Bisazza, A.; Miletto Petrazzini, M.E.; Giurfa, M. Honeybees use absolute rather than relative numerosity in number discrimination. *Biol. Lett.* **2019**, *15*, 20190138. [CrossRef]
10. Adámková, J.; Svoboda, J.; Benediktová, K.; Martini, S.; Nováková, P.; Tůma, D.; Burda, H. Directional preference in dogs: Laterality and "pull of the north". *PLoS ONE* **2017**, *12*, e0185243. [CrossRef]
11. Bisazza, A.; De Santi, A.; Bonso, S.; Sovrano, V.A. Frogs and toads in front of a mirror: Lateralisation of response to social stimuli in tadpoles of five anuran species. *Behav. Brain Res.* **2002**, *134*, 417–424. [CrossRef]
12. Santacà, M.; Agrillo, C. Two halves are less than the whole: Evidence of a length bisection bias in fish (*Poecilia reticulata*). *PLoS ONE* **2020**, *15*, e0233157. [CrossRef]
13. Gierszewski, S.; Bleckmann, H.; Schluessel, V. Cognitive abilities in Malawi cichlids (*Pseudotropheus* sp.): Matching-to-sample and image/mirror-image discriminations. *PLoS ONE* **2013**, *8*, e57363. [CrossRef] [PubMed]
14. Kuba, M.J.; Byrne, R.A.; Burghardt, G.M. A new method for studying problem solving and tool use in stingrays (*Potamotrygon castexi*). *Anim. Cogn.* **2010**, *13*, 507–513. [CrossRef] [PubMed]
15. Lebovich, L.; Darshan, R.; Lavi, Y.; Hansel, D.; Loewenstein, Y. Idiosyncratic choice bias naturally emerges from intrinsic stochasticity in neuronal dynamics. *Nat. Hum. Behav.* **2019**, *3*, 1190–1202. [CrossRef]
16. Andrade, C.; Alwarshetty, M.; Sudha, S.J.; Suresh Chandra, J. Effect of innate direction bias on T-maze learning in rats: Implications for research. *J. Neurosci. Methods* **2001**, *110*, 31–35. [CrossRef]
17. Danisman, E.; Bshary, R.; Bergmüller, R. Do cleaner fish learn to feed against their preference in a reverse reward contingency task? *Anim. Cogn.* **2010**, *13*, 41–49. [CrossRef] [PubMed]
18. Agrillo, C.; Parrish, A.E.; Beran, M.J. Do primates see the solitaire illusion differently? A comparative assessment of humans (*Homo sapiens*), chimpanzees (*Pan troglodytes*), rhesus mon-keys (*Macaca mulatta*), and capuchin monkeys (*Cebus apella*). *J. Comp. Psychol.* **2014**, *128*, 402–413. [CrossRef]
19. Alves, C.; Chichery, R.; Boal, J.G.; Dickel, L. Orientation in the cuttlefish *Sepia officinalis*: Response versus place learning. *Anim. Cogn.* **2007**, *10*, 29–36. [CrossRef]
20. Miletto Petrazzini, M.E.; Wynne, C.D. What counts for dogs (*Canis lupus familiaris*) in a quantity discrimination task? *Behav. Proc.* **2016**, *122*, 90–97. [CrossRef]
21. Miletto Petrazzini, M.E.; Mantese, F.; Prato Previde, E. Food quantity discrimination in puppies (*Canis lupus familiaris*). *Anim. Cogn.* **2020**, *23*, 703–710. [CrossRef] [PubMed]
22. Collins, R.L. When left-handed mice live in right-handed world. *Science* **1975**, *187*, 181–184. [CrossRef] [PubMed]
23. Ayroles, J.F.; Buchanan, S.M.; O'Leary, C.; Skutt-Kakaria, K.; Grenier, J.K.; Clark, A.G.; Hartl, D.L.; de Bivort, B.L. Behavioral idiosyncrasy reveals genetic control of phenotypic variability. *Proc. Natl. Acad. Sci. USA* **2015**, *112*, 6706–6711. [CrossRef] [PubMed]
24. Iigaya, K.; Ahmadian, Y.; Sugrue, L.P.; Corrado, G.S.; Loewenstein, Y.; Newsome, W.T.; Fusi, S. Deviation from the matching law reflects an optimal strategy involving learning over multiple timescales. *Nat. Commun.* **2019**, *10*, 1466. [CrossRef]

25. Frasnelli, E.; Ponte, G.; Vallortigara, G.; Fiorito, G. Visual lateralization in the cephalopod mollusk *Octopus vulgaris*. *Symmetry* **2019**, *11*, 1121. [CrossRef]
26. Regaiolli, B.; Spiezo, C.; Hopkins, W.D. Asymmetries in mother-infant behaviour in Barbary macaques (*Macaca sylvanus*). *PeerJ* **2018**, *6*, e4736. [CrossRef]
27. Dadda, M.; Agrillo, C.; Bisazza, A.; Brown, C. Laterality enhances numerical skills in the guppy, *Poecilia reticulata*. *Front. Behav. Neurosci.* **2015**, *9*, 285. [CrossRef]
28. Bisazza, A.; Rogers, L.J.; Vallortigara, G. The origins of cerebral asymmetry: A review of evidence of behavioural and brain lateralization in fishes, reptiles and amphibians. *Neurosci. Biobehav. Rev.* **1998**, *22*, 411–426. [CrossRef]
29. Güntürkün, O.; Ströckens, F.; Ocklenburg, S. Brain lateralization: A comparative perspective. *Physiol. Rev.* **2020**, *100*, 1019–1063. [CrossRef]
30. Ströckens, F.; Güntürkün, O.; Ocklenburg, S. Limb preferences in non-human vertebrates. *Laterality* **2013**, *18*, 536–575. [CrossRef]
31. Versace, E.; Caffini, M.; Werkhoven, Z.; de Bivort, B. Individual, but not population asymmetries, are modulated by social environment and genotype in *Drosophila melanogaster*. *Sci. Rep.* **2020**, *10*, 4480. [CrossRef] [PubMed]
32. Miletto Petrazzini, M.E.; Sovrano, V.; Vallortigara, G.; Messina, A. Brain and behavioral asymmetry: A lesson from fish. *Front. Neuroanat.* **2020**, *14*, 11. [CrossRef] [PubMed]
33. Dadda, M.; Zandonà, E.; Agrillo, C.; Bisazza, A. The costs of hemispheric specialization in a fish. *Proc. R. Soc. B Biol. Sci.* **2009**, *276*, 4399–4407. [CrossRef] [PubMed]
34. Bandmann, O.; Burton, E.A. Genetic zebrafish models of neurodege-nerative diseases. *Neurobiol. Dis.* **2010**, *40*, 58–65. [CrossRef] [PubMed]
35. Langenau, D.M.; Zon, L.I. The zebrafish: A new model of T-cell and thymic development. *Nat. Rev.* **2005**, *5*, 307–317. [CrossRef] [PubMed]
36. Agrillo, C.; Miletto Petrazzini, M.E.; Tagliapietra, C.; Bisazza, A. Inter-specific differences in numerical abilities among teleost fish. *Front. Psychol.* **2012**, *3*, 483. [CrossRef]
37. Potrich, D.; Sovrano, V.A.; Stancher, G., Vallortigara, G. Quantity discrimination by zebrafish (*Danio rerio*). *J. Comp. Psychol.* **2015**, *129*, 388–393. [CrossRef]
38. Potrich, D.; Rugani, R.; Sovrano, V.A.; Regolin, L.; Vallortigara, G. Use of numerical and spatial information in ordinal counting by zebrafish. *Sci. Rep.* **2019**, *9*, 18323. [CrossRef]
39. Baratti, G.; Potrich, D.; Sovrano, V.A. The environmental geometry in spatial learning by zebrafish (*Danio rerio*). *Zebrafish* **2020**, *17*, 131–138. [CrossRef]
40. Gori, S.; Agrillo, C.; Dadda, M.; Bisazza, A. Do fish perceive illusory motion. *Sci. Rep.* **2014**, *4*, 6443. [CrossRef]
41. Santacà, M.; Lucon-Xiccato, T.; Agrillo, C. The Delboeuf illusion's bias in food choice of teleost fishes: An interspecific study. *Anim. Behav.* **2020**, *164*, 105–112. [CrossRef]
42. Facchin, L.; Burgess, H.A.; Siddiqi, M.; Granato, M.; Halpern, M.E. Determining the function of zebrafish epithalamic asymmetry. *Philos. Trans. R. Soc. B Biol. Sci.* **2009**, *364*, 1021–1032. [CrossRef] [PubMed]
43. Stennett, C.R.; Strauss, R.E. Behavioural lateralization in zebrafish and four related species of minnows (Osteichthyes: *Cyprinidae*). *Anim. Behav.* **2010**, *79*, 1339–1342. [CrossRef]
44. Hata, H.; Hori, M. Inheritance patterns of morphological laterality in mouth opening of zebrafish, *Danio rerio*. *Laterality* **2001**, *17*, 741–754. [CrossRef]
45. Brown, C.; Western, J.; Braithwaite, V.A. The influence of early experience on, and inheritance of cerebral lateralization. *Anim. Behav.* **2007**, *74*, 231–238. [CrossRef]
46. Domenici, P.; Allan, B.; McCormick, M.I.; Munday, P.L. Elevated carbon dioxide affects behavioural lateralization in a coral reef fish. *Biol. Lett.* **2012**, *8*, 78–81. [CrossRef]
47. Sovrano, V.A.; Bisazza, A.; Vallortigara, G. Modularity and spatial reorientation in a simple mind: Encoding of geometric and nongeometric properties of a spatial environment by fish. *Cognition* **2002**, *85*, B51–B59. [CrossRef]
48. Agrillo, C.; Miletto Petrazzini, M.E.; Bisazza, A. Numerical abilities in fish: A methodological review. *Behav. Proc.* **2007**, *141*, 161–171. [CrossRef]
49. Oliveira, J.; Silveira, M.; Chacon, D.; Luchiari, A. The zebrafish world of colors and shapes: Preference and discrimination. *Zebrafish* **2015**, *12*, 166–173. [CrossRef]

50. Savage, A.; Dronzek, L.A.; Snowden, C.T. Color discrimination by the cotton-top tamarin (*Saguinus oedipus oedipus*) and its relation to fruit coloration. *Folia Primatol.* **1987**, *49*, 57–69. [CrossRef]
51. Giurfa, M. Conditioning procedure and color discrimination in the honeybee *Apis mellifera*. *Naturwissenschaften* **2004**, *91*, 228–231. [CrossRef] [PubMed]
52. Dunn, T.W.; Fitzgerald, J.E. Correcting for physical distortions in visual stimuli improves reproducibility in zebrafish neuroscience. *eLife* **2020**, *9*, e53684. [CrossRef] [PubMed]
53. Uller, C.; Lewis, J. Horses (*Equus caballus*) select the greater of two quantities in small numerical contrasts. *Anim. Cogn.* **2009**, *12*, 733–738. [CrossRef] [PubMed]
54. Bánszegi, O.; Urrutia, A.; Szenczi, P.; Hudson, R. More or less: Spontaneous quantity discrimination in the domestic cat. *Anim. Cogn.* **2016**, *19*, 879–888. [CrossRef] [PubMed]
55. Uller, C.; Jaeger, R.; Guidry, G.; Martin, C. Salamanders (*Plethodon cinereus*) go for more: Rudiments of number in an amphibian. *Anim. Cogn.* **2003**, *6*, 105–112. [CrossRef]
56. Pritchard, V.L.; Lawrence, J.; Butlin, R.K.; Krause, J. Shoal size in zebrafish, *Danio rerio*: The influence of shoal size and activity. *Anim. Behav.* **2001**, *62*, 1085–1088. [CrossRef]
57. Dadda, M.; Bisazza, A. Does brain asymmetry allow efficient performance of two simultaneous tasks? *Anim. Behav.* **2006**, *72*, 523–529. [CrossRef]
58. Gatto, E.; Agrillo, C.; Brown, C.; Dadda, M. Individual differences in numerical skills are influenced by brain lateralization in guppies (*Poecilia reticulata*). *Intelligence* **2019**, *74*, 12–17. [CrossRef]
59. Gatto, E.; Lucon-Xiccato, T.; Bisazza, A.; Manabe, K.; Dadda, M. The devil is in the detail: Zebrafish learn to discriminate visual stimuli only if salient. *Behav. Proc.*. under review.
60. Bisazza, A.; Agrillo, C.; Lucon-Xiccato, T. Extensive training extends numerical abilities of guppies. *Anim. Cogn.* **2014**, *7*, 1413–1419. [CrossRef]
61. Sovrano, V.A.; Bisazza, A. Perception of subjective contours in fish. *Perception* **2009**, *38*, 579–590. [CrossRef] [PubMed]
62. Miletto Petrazzini, M.E.; Agrillo, C.; Izard, V.; Bisazza, A. Do humans (*Homo sapiens*) and fish (*Pterophyllum scalare*) make similar numerosity judgments? *J. Comp. Psychol.* **2016**, *130*, 380–390. [CrossRef] [PubMed]

MDPI
St. Alban-Anlage 66
4052 Basel
Switzerland
Tel. +41 61 683 77 34
Fax +41 61 302 89 18
www.mdpi.com

Symmetry Editorial Office
E-mail: symmetry@mdpi.com
www.mdpi.com/journal/symmetry

www.ingramcontent.com/pod-product-compliance
Lightning Source LLC
LaVergne TN
LVHW070609170726
843515LV00004B/25

* 9 7 8 3 0 3 6 5 0 6 1 2 8 *